Y0-BCL-952

BOOK ORDER INFORMATION
REPAIRMASTER
P.O. BOX 649
WEST JORDAN, UT 84084
BANK CARD ORDERS 1-800-347-5163

VOLUME 16 7551 ISBN 1-56302-045-9

32 VOLUME SET ISBN 1-56302-105-6

EDITORIAL STAFF

Editor/Tech Writer .... *Barnee Schollnick*
Director .... *Woody Wooldridge*

REPAIR-MASTER for..
DOMESTIC
REFRIGERATION

PRINTED IN U.S.A.

33401088

2-96

FOREWORD

This Repair Master contains information and service procedures to assist the service technician in correcting conditions that are not always obvious.

A thorough knowledge of the functional operation of the many component parts used on appliances is important to the serviceman, if he is to make a proper diagnosis when a malfunction of any part occurs.

We have used many representative illustrations, diagrams and photographs to portray more clearly these various components for a better over-all understanding of their use and operation.

IMPORTANT SAFETY NOTICE

You should be aware that all major appliances are complex electromechanical devices. Master Publication's REPAIR MASTER® Service Publications are intended for use by individuals possessing adequate backgrounds of electronic, electrical and mechanical experience. Any attempt to repair a major appliance may result in personal injury and property damage. Master Publications cannot be responsible for the interpretation of its service publications, nor can it assume any libility in connection with their use.

SAFE SERVICING PRACTICES

To preclude the possibility of resultant personal injury in the form of electrical shock, cuts, abrasions or burns, etc., that can occur spontaneously to the individual while attempting to repair or service the appliance; or may occur at a later time to any individual in the household who may come in contact with the appliance, Safe Servicing Practices must be observed. Also property damage, resulting from fire, flood, etc., can occur immediately or at a later time as a result of attempting to repair or service — unless safe service practices are observed.

The following are examples, but without limitation, of such safe practices:

1. Before servicing, always disconnect the source of electrical power to the appliance by removing the product's electrical plug from the wall receptacle, or by removing the fuse or tripping the circuit breaker to OFF in the branch circuit servicing the product.

NOTE: If a specific diagnostic check requires electrical power to be applied such as for a voltage or amperage measurements, reconnect electrical power only for time required for specific check, and disconnect power immediately thereafter. During any such check, ensure no other conductive parts, panels or yourself come into contact with any exposed current carrying metal parts.

2. Never bypass or interfere with the proper operation of any feature, part, or device engineered into the appliance.

3. If a replacement part is required, use the specified manufacturers part, or an equivalent which will provide comparable performance.

4. Before reconnecting the electrical power service to the appliance — be sure that:

    a. All electrical connections within the appliance are correctly and securely connected.
    b. All electrical harness leads are properly dressed and secured away from sharp edges, high-temperature components such as resistors, heaters, etc., and moving parts.
    c. Any uninsulated current-carrying metal parts are secured and spaced adequately from all non-current carrying metal parts.
    d. All electrical ground, both external and internal to the product are correctly and securely connected.
    e. All water connections are properly tightened.
    f. All panels and covers are properly and securely reassembled.

5. Do not attempt an appliance repair if you have any doubts as to your ability to complete it in a safe and satisfactory manner.

MASTER PUBLICATIONS

TABLE OF CONTENTS

TABLE OF CONTENTS

SECTION 1

SERVICE CHECK LIST

The following diagnosis chart is intended to be only a starting point in proceeding with the servicing of refrigerators. The diagnosis chart can only deal in generalities; to effectively service any appliance, the serviceman must thoroughly understand the mechanical functions and electrical circuitry of the appliance.

A considerable amount of time and money can be saved if a serviceman will take time to analyze the probable cause of a malfunction of a machine before proceeding to remove any parts. Always be sure first that the machine is properly installed and its power cord is plugged into a live receptacle that is properly fused. Check also, that the thermostat is properly set and that food is placed on the refrigerator shelves in such a manner to allow proper air flow. Check to be sure the condenser is clean and has enough clearance for a good flow of air across it. Make sure the door gasket is sealing properly and that the interior light is not staying on.

Always make a visual check first before using any testing equipment such as test lamps, voltmeters or ohmmeters. Before attempting to remove any electrical part from the machine, disconnect the power cord from the live receptacle. If a voltmeter or test lamp is being used for testing, the power cord must be plugged into a live receptacle, however.

CONDITION	POSSIBLE CAUSE	REMEDY
UNIT WILL NOT RUN	Inoperative thermostat	Replace.
	Service cord pulled out of wall receptacle	Replace.
	Service cord pulled out of bulk-head	Reconnect.
	No voltage at wall receptacle (house fuse blown).	Replace fuse.
	Faulty cabinet wiring	Repair as required.
	Relay cord pulled out of bulk-head	Reconnect.
	Relay loose or inoperative	Direct-test compressor using test cord. If unit now runs, replace relay.
	Compressor windings open	Replace compressor.
	Low voltage causing compressor to cycle on overload. (Voltage fluctuation should not exceed 10% plus or minus, from nominal rating 115 volts.)	Check for cause of low voltage and correct.
UNIT RUNS – BUT NO REFRIGERATION	System out of refrigerant	Check for leaks and repair as necessary.
	Compressor not pumping	Replace.
	Restricted strainer	Replace.
	Restricted filter drier	Replace.
	Restricted capillary tube	Replace.
	Moisture in system	Evacuate, install new dryer and recharge system.
	Defrost solenoid valve stuck open	Feel by-pass line for excessive temperature. Check defrost control. Repair as required.

CONDITION	POSSIBLE CAUSE	REMEDY
UNIT SHORT CYCLES	Erratic thermostat	Replace.
	Condenser fan not running	Replace (forced convection models).
	Faulty relay	Replace.
	Restricted air flow over condenser (static condenser models)	Relocate cabinet.
	Low voltage	Check for cause and correct.
	Condenser fan shroud not in place. (Forced convection models)	Shroud must be in place for proper air flow across condenser.
	Dirty condenser	Clean.
	Inoperative provision compartment fan	Replace.
	Inoperative freezer compartment fan	Replace.
	Compressor draws excessive wattage	Check with wattmeter and replace if defective.
UNIT RUNS TOO MUCH, OR 100%	Erratic thermostat or thermostat set too cold	Replace or reset to normal position
	Dirty condenser (forced convection models)	Clean.
	Condenser fan shroud not in place (forced convection models)	Shroud must be in place for proper air flow.
	Condenser fan not operating	Replace (Forced convection models)
	Refrigerator exposed to unusual heat or abnormally high room temperature	Relocate or ventilate room.
	Inefficient compressor	Replace.
	Door gaskets not sealing	Check and adjust hinges and strike or replace gasket if necessary.
	System undercharged or overcharged	Correct charge.
	Interior light stays on	Check door switch.
	Non-condensables in system	Evacuate and recharge.

CONDITION	POSSIBLE CAUSE	REMEDY
UNIT RUNS TOO MUCH, OR 100% (Continued)	Defrost solenoid valve bypassing refrigerant	Flush valve or replace.
	Capillary tube kinked or partially restricted.	Replace or clean out with cap tube cleaner.
	Filter drier or strainer partially restricted	Replace.
	Food compartment liner heater grounded to liner	Replace.
UNIT NOISY	Tubing vibrates	Adjust tubing.
	Internal compressor noise	Replace compressor.
	Condenser fan blade loose or bent	Repair.
	Condenser fan blade hitting shroud	Adjust shroud.
	Compressor vibrating on cabinet frame	Adjust.
	Loose water evaporating pan	Reposition.
	Rear machine compartment baffle missing	Replace.
	Compressor operating at high head pressure due to dirty condenser	Clean condenser.
	Compressor operating at high head pressure due to condenser fan not running	Replace fan motor.
	Needs sound deadener kit	Install.
	Compressor needs damping strap assembly	Install.
FREEZER COMPARTMENT TOO WARM	Erratic thermostat or thermostat set too warm	Adjust or replace.
	Freezer compartment door left open	
	Freezer compartment gaskets not sealing	Check and adjust hinges and strike or replace.

CONDITION	POSSIBLE CAUSE	REMEDY
FREEZER COMPARTMENT TOO WARM (Continued)	Inoperative or erratic operating freezer compartment fan motor	Replace.
	Improperly positioned freezer compartment fan blade	Reposition properly.
	Inoperative or erratic operating freezer compartment door switch	Replace.
	Freezer compartment evaporator coil iced up	Check conditions under Freezer Coil Blocked With Ice.
	Inoperative or erratic operating automatic defrost control	Replace.
	Defrost valve solenoid burned out	Replace.
	Wire loose at automatic defrost control or at solenoid of defrost valve	Reconnect Wire.
	Wire freezer rack reversed, restricting air flow	Position rack properly.
	Excessive freezer compartment service load	Customer education.
	Hot-gas bypass line restricted	Replace.
	Drain trough heater burned out or wire off	Replace.
	Abnormally low room temperatures	Relocate cabinet or install mullion heater or condenser.
FREEZER COIL BLOCKED WITH ICE	Inoperative or erratic automatic defrost control	Replace.
	Defrost control termination too low	Check and correct as necessary.
	Defrost control thermal element not properly installed	Install correctly.
	Defrost control thermal element broken	Replace defrost control.
	Defrost control incorrectly wired	Check wiring.

CONDITION	POSSIBLE CAUSE	REMEDY
FREEZER COIL BLOCKED WITH ICE (Continued)	Inoperative freezer compartment door switch or freezer compartment fan motor	Replace.
	Defrost valve solenoid burned out or loose wire	Replace solenoid or connect wire.
	Defrost valve stuck in closed position	Flush valve or replace.
	Hot-gas by-pass line restricted	Replace.
	Freezer compartment drain plugged	Clean.
	Freezer compartment drain sump or drain trough heater burned out	Replace.
REFRIGERATOR STAYS IN DEFROST CYCLE	Defrost solenoid valve stuck open	Flush valve or replace.
	Defrost control termination temperature too high	Check and correct as necessary.
	Defrost control incorrectly wired	Check wiring.
	Inoperative automatic defrost control	Replace.
	Abnormally low room temperature (below 55°)	Relocate cabinet or provide heat in room.
FOOD COMPARTMENT TOO WARM	Thermostat erratic, or cut-in and cut-out too high	Replace if erratic or recalibrate.
	Incorrect thermostat installed	Replace.
	Thermostat thermal element not installed properly in thermal well.	Install properly.
	System short of refrigerant	Correct charge. Check for leaks.
	Inefficient compressor	Replace.
	Dirty condenser (forced convection models)	Clean.
	Condenser fan not operating	Replace fan motor.

CONDITION	POSSIBLE CAUSE	REMEDY
FOOD COMPARTMENT TOO WARM (Continued)	Inoperative food compartment fan motor	Replace.
	Food compartment fan blade improperly positioned	Adjust.
	Inoperative food compartment door switch	Replace.
	Inoperative freezer compartment fan motor	Replace.
	Inoperative freezer compartment door switch	Replace.
	Door gasket not sealing	Adjust or replace.
	Excessive service load	Inform customer.
	Evaporator baffle not installed properly	Install properly.
	Restricted strainer, filter drier, or capillary tube	Replace.
	Shelves covered with foil wrap or paper, retarding air circulation	Remove foil and instruct customer.
	Restricted air flow over condenser (static condenser models)	Remove restriction.
	Freezer coil blocked with ice	See symptoms under this heading.
WATER IN BOTTOM OF FOOD COMPARTMENT	Drain tube frozen shut or plugged	Clean.
	Evaporator baffle not installed properly	Install properly.
	Humidiplate warped forward	Adjust.
	Drain trough split	Replace.
	Water draining between trough and liner wall	Seal with Permagum.
	Food compartment liner bowed	Replace.

CONDITION	POSSIBLE CAUSE	REMEDY
WATER IN BOTTOM OF FOOD COMPARTMENT (Continued)	Door gasket not sealing properly	Adjust or replace.
WATER OR ICE IN BOTTOM OF FREEZER COMPARTMENT	Drain tube frozen shut or plugged	Clean.
	Evaporator cover plate not in place	Install properly.
	Freezer sump heater burned out	Replace.
	Drain trough heater burned out	Replace.
CONDENSATION AROUND DOORS	Door gaskets not sealing properly	Adjust or replace.
	Mullion drier coil burned out or wire lead off	Replace coil or reconnect wire lead
	Abnormally high humidity	Customer education

WHAT IS HEAT?

In refrigeration, and in air conditioning as well, we are concerned primarily with the transfer of heat. We do not actually make things cold. We merely remove heat from them. By the same token, when we make things warm or hot, we add heat.

To understand refrigeration, therefore, we must know something about heat. Notice that we have said nothing about cold. Actually, cold is an abstract concept, not a specific concrete thing. Cold is really a relative term meaning that there is little or no heat. A sort of rough and ready definition of cold would be to say that cold is the lack of heat, in the same way that darkness is really the lack of light and is not a physical thing in itself.

SENSIBLE HEAT AND LATENT HEAT

Heat is a form of energy just as electricity is a form of energy. Heat can, therefore, do work. It exists in two forms - sensible heat and latent heat. Sensible heat is the kind of heat with which we are most familiar because it affects our senses. In other words, we can feel it and measure its intensity. If we touch a hot object, it feels hot and we can measure the intensity of this heat by means of a thermometer. Latent heat, on the other hand, is not so easily recognized, although it is no less real. We will discuss latent heat a little later in this section; but for the time being, let us confine our discussion to sensible heat.

TEMPERATURE

Most people think that temperature is an indication of the quantity of heat contained in a substance. This is incorrect. Temperature is, instead, an indication of *the intensity of the sensible heat in the substance*. Note two things about that definition: (1) Temperature does not indicate the total quantity of heat, but only its intensity. We can, therefore, have a small amount of heat at a high temperature if it is concentrated enough, or we can have a very large amount of heat at a low temperature if it is not concentrated. (2) The second important fact we must note is that latent heat is not measured as temperature. Since we do not as yet know what latent heat is, we will not discuss the second point here, but we should make a mental note of this fact because it will be referred to later.

We can think of temperature in this way. Suppose we had some red dye in an eye dropper and two containers of water, as shown in *Figure 1*, one holding a point, the other holding a gallon. If we add one drop of the red dye to each container, the pint will become much redder than the gallon. Yet, the actual amount of dye added is the same in both cases. We can consider the red dye as sensible heat and the color as temperature. The same amount of red dye added to the smaller quantity of water will produce a deeper shade of red. In other words, the intensity of color will be greater, or the temperature will be higher.

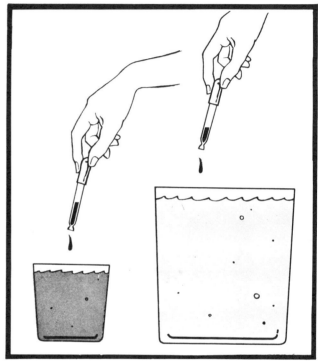

Figure 1

The instrument used to indicate temperature is called a thermometer. A thermometer will tell us how hot a substance is, but will not tell us how much actual heat the substance contains. In other words, a thermometer measures the heat intensity, not the total heat.

We often say that a thermometer tells us how hot something is, and we will use the words hot and cold throughout this course of study. It is, therefore, important for us to bear in mind that the word "hot" does not refer to the amount of heat contained in a substance, but rather to its temperature, by which we mean the intensity of its sensible heat.

In the English system, which is also used in the United States, temperature is measured according to the Fahrenheit scale. On this scale, the freezing point of water is 32° (also written 32°F.), and the boiling point of water is 212°F. Another commonly used scale of temperature is the Centigrade scale. In the Centigrade system, the freezing point of water is 0° (also written 0°C.), and the boiling point of water is 100°C.

There is no upper limit to temperature. In other words, the temperature of a substance can be increased indefinitely. However, there is a definite limit to how cold a substance can be made. This lower limit is −460°F. on the Fahrenheit scale. When a substance is cooled down to this temperature, all heat has been removed. We call this temperature absolute zero. While we have never cooled anything down to absolute zero, we have come very close to it.

MEASURING HEAT
As we have seen, temperature is not a measurement of heat. We must, therefore, have some other unit of heat if we are to work with it. In the English system, the unit of heat is called the British Thermal Unit, which is abbreviated B.T.U. One B.T.U. is defined as the amount of heat required to raise the temperature of 1 lb. of water 1°F.

If we raise the temperature of 1 pt. of water 1°F., we have added 1 B.T.U. This is because 1 pt. of water weighs 16 ozs. or 1 lb. By the same token, if we raise the temperature of 1 gal. (8 lbs.) of water 1° F., we have added 8 B.T.U.

FLOW OF HEAT
Heat always flows from a hotter to a colder substance. If we touch a cold object to a hot one the heat will flow from the hot to the cold object, and eventually enough heat will be transferred so that the temperatures of both objects will be equal. When that happens, the flow of heat stops.

Note that the heat flow is always from the hotter to the colder substance regardless of the actual amount of heat present in either substance. It is the intensity of heat (that is, the temperature) which determines the direction of heat flow. For example, if you put your cold hands on a hot radiator, as shown in *Figure 2*, heat will move from the radiator to your hands and warm them. Heat will never flow from your hands to the warm radiator. If it did, your hands would get cold rather than warm, and you know that this will never happen. Similarly, if you pick up an ice cube, your hands will immediately start to feel cold and the ice cube will melt as it absorbs heat. Your hand

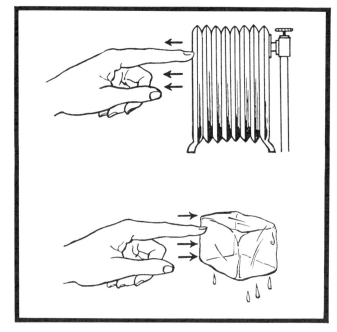

Figure 2

will never get warm by touching an ice cube.

HOW SENSIBLE HEAT IS TRANSFERRED
Sensible heat may be transferred from one body to another in three ways. These are:

1. Conduction
2. Convection
3. Radiation

Conduction
Conduction, as we use it here, means contact. When we heat one end of a metal bar, for example, the molecules are speeded up at the point where heat is applied. These speedier molecules, in turn, collide with and speed up the adjoining molecules, thus raising the temperature of the metal adjoining the heated portion. Heat thus flows along the metal bar.

Similarly, if we place a hot object in contact with a cold one, the molecules of the hot object will be close enough to the molecules of the cold one so that they can bump into and speed them up, thus raising their temperature.

There is a wide variation in the ease with which heat can be transferred or passed along by conduction. Some materials, principally metals, are good conductors of heat, while others are poor. In a general way, you will find that materials which are good conductors of electricity are also good conductors of heat. Gases are poor heat conductors. Most insulating materials used in refrigeration are very light and porous, and it is the air contained in the pores of the material which provides most of the insulating value.

Convection

Heat transfer by convection refers primarily to fluids. In this connection, we must remind you that a fluid is something which flows, so the term applies both to gases and to liquids.

Convection arises from the fact that when a substance is heated, it generally expands and becomes lighter; when it is cooled, it contracts and becomes heavier. If, therefore, we place a pan of water over a gas flame, as shown in *Figure 3* the bottom of the pan will heat up by conduction, since it is in contact with the hot flame. The pan, in turn, also heats up the layer of water contacting it by conduction. This heated water expands and becomes lighter, so that it rises. Cold, heavier water flows in to take its place and, in turn, is heated. This sets up a regular circulation pattern, which continues as long as heat is applied. This flow is termed convection. In a refrigerator, the food compartment is cooled by the evaporator

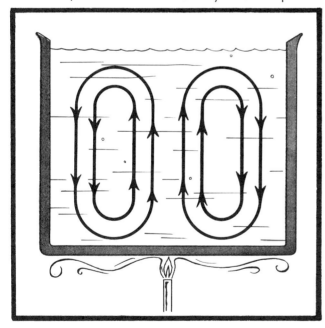

Figure 3

through the process of convection, as shown in *Figure 4*. The cold evaporator is always at the top of the food compartment. When the air in contact with it gets cold, this cold air contracts, becomes heavier and sinks to the bottom of the food compartment. Air from the sides comes up to replace the cooled air, becomes cold, in turn, and sinks. In this way, circulation of cold air is maintained. As the cold air contacts the warmer food, it picks up heat from the food by conduction. This,

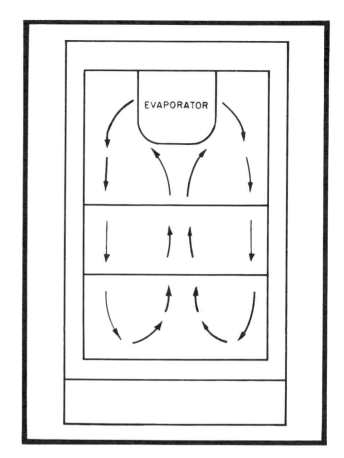

Figure 4

of course, warms the air and thus adds to the strength of the convection currents.

Radiation

When we stand in front of a fire, we feel its warmth. Have you ever wondered how, or even why, the heat of the fire reaches you? It does not do so by conduction because you are not in contact with the fire. It cannot do so by convection because the heated air above the fire rises and the colder air from the sides moves in to take its place. In other words, the convection currents flow from you to the fire, not from the fire to you. What then is left? The answer is: radiation.

All bodies containing heat, no matter how small an amount, radiate some of that heat in all directions, in exactly the same way as a lighted lamp radiates light. As a matter of fact, these heat rays are of exactly the same nature as light rays, except that their wave length is somewhat longer. Radio waves, heat rays, and light rays are all electromagnetic waves or radiations, the primary difference among them being in the length of their individual waves. When a radio wave strikes a good conductor of electricity, such as metal, the waves are absorbed and the energy they contain is converted into electrical energy, so that an electric current flows through the conductor. It is this current which is amplified and heard in radio and TV receivers.

When heat or light radiations strike any body which absorbs them, their energy is converted into sensible heat. If these radiations strike something which is transparent to them, they simply pass through with no effect. If we hold a sheet of glass in front of a fire, the sheet of glass will not warm up appreciably because both the light and the radiant heat pass through. However, if we hold a sheet of metal in front of the fire, the metal will rapidly heat up, because the heat and light rays are absorbed and converted into sensible heat. In this connection, we should point out that the heat rays, which have the longer wave length, contain most of the energy and, therefore, mainly responsible for the heating.

Light, which has the shorter wave length, appears red to the human eye. Because heat rays are even longer than red light waves, they are called infrared rays, or infrared light.

The amount of infrared energy radiated by a body depends upon its temperature—the higher the temperature, the greater the radiation. As an example, you are constantly radiating heat to everything around you, but at the same time, you are absorbing heat from their radiation to you. If you are close to a warm surface, you receive more heat from it than you give to it, so it feels warm. On the other hand, if you are close to a cold body, you will radiate more heat to it than it will to you and you will feel cold. For example, if you are in a poorly insulated room in cold weather in which the walls are colder than the air in the room, you will radiate more heat to the walls than they will radiate to you. As a result, you will feel colder than you would feel on a day when the walls are warmer, even though the actual air temperature within the room may be exactly the same.

Radiation always takes place from the surface of the radiating body. Surfaces differ considerably in their ability to radiate heat energy. Bright, shiny surfaces, such as highly polished metals, make poor radiators; dull, dark surfaces make good radiators. It so happens that surfaces which absorb radiations well also radiate well, and surfaces which absorb radiations poorly (in other words, are good reflectors) also make poor radiators.

That is why condensers in refrigeration systems are always painted in dull black, since the purpose of the condenser is to lose as much heat as possible. On the other hand, the glass liner of a vacuum type thermos bottle is silvered so as to retard heat flow by radiation as much as possible.

SPECIFIC HEAT

A B.T.U. is defined as the amount of heat required to raise the temperature of 1 lb. of water 1°F. Actually, this is not perfectly true, since the amount of heat required varies slightly, depending upon the temperature of the water. The standard B.T.U. is the average for a temperature range from 32°F. to 212°F. at an atmospheric pressure of 14.7 pounds per square inch.

If we measured the amount of heat required to raise the temperature of 1 lb. of ice 1°F., we would find it to be only half as much, or .5 B.T.U. Similarly, 1 lb. of air would require only .24 B.T.U. to raise its temperature 1°F. We say that the specific heat of water is 1, the specific heat of ice .5, and the specific heat of air is .24.

The specific heats of some common substances are given in Table 1.

SUBSTANCE	SPECIFIC HEAT
Water	1.00
Corkboard	.48
Glass	.19
Ice	.5
Rubber	.48
Air	.24
Aluminum	.219
Brass	.090
Copper	.093
Iron	.119
Lead	.031

The refrigeration engineer, who may be called on to design refrigeration systems, must know the specific heats of the various products which his equipment is intended to refrigerate so that he can calculate the amount of refrigerating capacity he will have to supply, and design his equipment accordingly.

LATENT HEAT

We add ice cubes to warm drinks in summer in order to cool them. Have you ever stopped to wonder why we do not add cold water instead? After all, there is not much difference in temperature between an ice cube at 32°F. and freezing cold water.

Let us suppose we had six ounces of water in a drinking glass at a temperature of 75°F., and we added one ounce of ice cold water, at say, 33°F., the heat would flow from the warm water to the cold until a balance was reached. The final temperature would be an average of the two, and if we would care to carry out the experiment, we would find that the temperature would be 69°F. in this instance. This is still not very cold.

Now let us try something else. Instead of adding one ounce of ice cold water we would add an ice cube weighing one ounce. We could easily get the temperature of the water in this experiment down to 50°F. and still have some ice left over. Obviously, there must be some factor besides the temperature which must be taken into consideration. This factor is latent heat.

We can show this latent heat in still another way. If we take the temperature of a large block of ice, we will find it to be 32°F. Let this block of ice remain in a warm room and we will take its temperature at intervals as it melts. We will find that the ice will remain at 32°F. until it is all gone, even though it has been absorbing large quantities of heat from the surrounding air. Obviously, the heat caused the ice to melt, but it did not raise the temperature of the ice. The heat was, somehow, absorbed by the ice in the melting process. This absorbed heat is called latent or hidden heat.

Now for another experiment, let us pour some cold water into a saucepan and insert a thermometer. Say the temperature of the water is 70°F. We will put the saucepan of water on a stove and turn on the heat. The temperature of the water will begin to rise. We can watch the rise by looking at the thermometer. This experiment illustrates the fact that adding heat to a substance such as water raises its temperature.

If we keep on adding heat, the temperature will continue rising until the water starts to boil. At this point, the conditions will change radically. If we watch the thermometer as in *Figure 5*, we will see that it registers 212°F. and that it will not rise above this temperature, no matter how much heat is added. From this point on, the addition of heat merely causes the water to boil away, but it does not raise the temperature of the water.

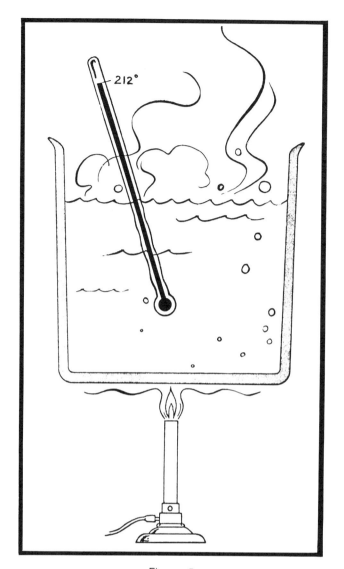

Figure 5

Furthermore, if we would hold the thermometer a little above the boiling water and measure the temperature of the escaping steam, we would find that its temperature would also be 212°F., as shown in *Figure 6*

From this we see that adding heat to a liquid raises its temperature only up to the point at which the liquid starts to boil. From that point on, adding heat merely changes the liquid into a gas or into a vapor. Furthermore, since the temperature of the vapor is the same as the temperature of the boiling liquid, all the heat we have added has been absorbed by the vapor without raising its temperature. This absorbed heat is latent heat.

The heat required to convert a solid into a liquid is called latent heat of fusion. The heat required to convert a liquid into a gas (vapor) is called latent heat of vaporization.

The amount of heat required to convert a specific quantity of ice into water is fixed and does not change. That heat is locked within the water itself and cannot escape as long as the water remains unchanged. However, if the water becomes ice, all the latent heat of fusion is given up. That is why an ice cube provides so much more cooling effect in our warm drinks than an equal quantity of water at the same temperature.

To melt one pound of ice at 32°F. without increasing its temperature requires 144 B.T.U. This same amount of heat added to water at 32°F. would increase the temperature to 176°F. You can

Figure 6

see, therefore, that the latent heat of fusion is quite a large quantity.

The latent heat of vaporization (which is the amount of heat required to convert water into water vapor) equals 970 B.T.U. In other words, it takes 970 B.T.U. to boil away one pound of water. We can see from this that the latent heat of vaporization is a tremendously large quantity, since it takes more than five times as much heat to boil away a pound of water as it does to raise its temperature from the freezing point to the boiling point. Bear this fact in mind. It is very important because all refrigeration systems operate on the principle of latent heat of vaporization. We must understand this principle before we will be able to understand how a refrigeration system works.

If latent heat still seems to be mysterious, let us look at it in this way. Suppose we had a block of steel on the floor and we lifted it up in the air. The block of steel would seem to be the same in all respects, but there is an important difference. We had to do work on the steel in order to lift it. Something was added to the steel, even though that something is not immediately apparent. Nevertheless, the work that went into lifting the steel is hidden within it in the form of energy. We may call this latent energy.

If we drop the steel, it will fall to the ground and in doing so it will give up the latent energy we put into it. If we wanted to do so, we could recover that energy in the form of work, and we would find that the energy recovered would be equal to the energy that went into the steel when it was lifted. In the same way, the latent heat used to heat a pound of water is exactly the same as the latent heat given up when the same quantity of water freezes.

THE MOLECULAR THEORY

Why does not latent heat show up as temperature in the same way that sensible heat does? The answer can be found if we look into the constitution of matter. If we could take a small quantity of any substance and keep dividing it into smaller and smaller particles, we would eventually reach the smallest particle of that substance. That smallest particle is the molecule. In other words, the smallest particle of water we can ever attain is a molecule of water. The smallest particle of iron we can ever attain is a molecule of iron, and so on for every substance.

It is possible to subdivide a molecule into even smaller particles, but when we do so we no longer have the same substance. For example, if we subdivide a molecule of water, we will get two particles of hydrogen and one particle of oxygen, but we will no longer have any water. The particles of hydrogen and oxygen are called atoms. In our study of refrigeration, we do not need to go beyond the definition of a molecule.

In any substance the molecules are held together by a strong attractive force, the nature of which is still unknown. If it were not for this force, the molecules would fly apart. Molecules are not still, but are darting about in a more or less random manner with tremendous velocities. This velocity of motion is really a measure or an indication of the temperature of the substance. The greater the temperature, the greater the velocity of motion. If the temperature is reduced, the motion is also reduced. If the temperature is reduced sufficiently, all motion will stop. The point at which this occurs is known as absolute zero because at this point there is no temperature and no heat. There can be no temperature below absolute zero.

In a solid, the molecules are closer together and their motion is more restricted than in either a liquid or a gas. If the temperature of the solid is increased, the motion of the molecules also increases. At some specific temperature, a sudden change occurs and the attractive bond between the molecules suddenly lessens. In short, the solid becomes a liquid. It takes energy to achieve this greater freedom of the molecules, and this energy must be supplied in the form of heat. This is the explanation of the latent heat of fusion. In other words, the molecules in the liquid state have a higher energy content than they did when they were in a solid state, just as the block of steel we referred to a little while back had a higher energy content when it was raised to a higher level. If more heat is added to the liquid, its molecules speed up with the absorption of this added heat, and the increase in molecular speed is indicated by a higher temperature. This addition continues until at another temperature the attractive force between the molecules is again greatly lessened and the molecules fly apart until they are separated by a much greater distance. At this point, the liquid has become a gas or a vapor. To achieve this increased molecular distance, energy in the form of heat was absorbed. This absorbed heat is the latent heat of vaporization.

SOLIDS, LIQUIDS AND GASES

Matter exists in three states. It may exist as a solid, as a liquid or as a gas. We are all familiar with these three states because under ordinary conditions, some substances are solids, others are liquids, and still others are gases.

We also know, in a general way, that many substances can exist in all of the three states, depending upon conditions. For example, water ordinarily exists as a liquid. However, if the temperature is lowered sufficiently, the water becomes a solid, and we call that solid ice. If the temperature of the water is raised, it becomes a gas, which we call steam.

In a solid, the motion of the molecules is restricted to such a degree that the substance will maintain its shape. For example, a cube of steel, or wood, or any other solid material, will retain its cubical shape indefinitely. Put a block of wood on a table top and the block of wood will remain a block of wood.

In a liquid, the molecules are much freer to move, and as a result, a liquid will not retain its shape. Pour a little water on a table, and the water will spread out in all directions. A liquid is fairly heavy, so that gravity will prevent it from rising, and as a result, the liquid will simply spread out until it forms a thin layer on any surface on which it is poured. If it is poured into a container, it will conform to the shape of the container, except that the top surface will be level because the force of gravity is uniform over the entire surface of the liquid.

In a gas, the molecules are spread so far apart and are so free to move that the force of gravity has very little effect. A gas will, therefore, expand in all directions. If we put a small quantity of a gas into a container, the gas will quickly fill the container uniformly. There will be no level above which there is no gas, as there is in the case of a liquid.

There are a good many substances which can exist in all three states, depending upon their temperature and the pressure exerted on them. Alcohol, for example, is a liquid at ordinary room temperature, but if we reduce the temperature sufficiently, the alcohol will freeze. The freezing point of grain alcohol is quite low, approximately −170°F. However, the boiling point of alcohol is above room temperature. This is why alcohol is ordinarily thought of as a liquid. The boiling point of alcohol is also fairly low, or 172°F., which is quite a bit lower than the 212°F. boiling point of water. Because of its low boiling point, alcohol evaporates quite rapidly and, of course, as it evaporates, it must absorb latent heat of vaporization from its surroundings.

This brings up a basic and important point. Whenever a liquid is changing into gas, whether due to boiling or to normal evaporation, it must absorb heat of vaporization. During slow evaporation, the absorption of such heat is quite slow, so we are not aware of it because the temperatures have plenty of time to equalize.

Remember that latent heat of vaporization is absorbed during boiling because the gas molecules have been given quite a lot of heat energy to give them this greater freedom of motion. The same transfer of heat energy must take place whether the gas is formed through rapid boiling or through slow evaporation.

If you pour a few drops of alcohol on the back of your hand, it will feel quite cool, even though the alcohol is at room temperature to begin with. This is because the alcohol starts to evaporate quickly as soon as it is out of the bottle, and in doing so, it absorbs sensible heat from everything around it. It absorbs sensible heat from the surrounding air, from your hand, and even from itself. This heat enters the alcohol as latent heat, but it is removed from your hand as sensible heat. You know by this time that when sensible heat is removed from a material, that material becomes colder. That is why the alcohol feels cold.

This is an important statement that must be understood if we are to understand the process of refrigeration. Therefore, let us repeat it: In order for a liquid to evaporate, it must absorb heat from the surroundings. If the evaporation is rapid enough, quite an appreciable lowering of the temperature may be achieved. In a modern refrigerator, a liquid, which is called the refrigerant, is caused to boil or evaporate at a rapid rate. This rapid evaporation absorbs so much heat that the food compartment of the refrigerator is cooled to something in the neighborhood of 40°F. The details of how the refrigerant is caused to evaporate in the refrigerator will be explained later.

ATMOSPHERIC PRESSURE

We live at the bottom of an ocean of air which extends above us for many miles. Even though air is very light, the many miles of air above us must necessarily weigh a great deal. How can we weigh it? The answer was found by an Italian physicist named Toricelli quite a number of years ago. He took a long glass tube about 36 inches in length and closed one end. He filled the tube with mercury and put his thumb over the open end so as to exclude all air. He then upended the tube, holding his thumb over the open end in order to prevent the mercury from spilling, and immersed the tube upright with the open end down in a dish of mercury. When he removed his thumb, he found that the mercury did not run out of the tube as he might have expected. Instead, the level dropped only slightly, as shown in *Figure 7* When he measured the height of the column of mercury, he found it to be 30 inches.

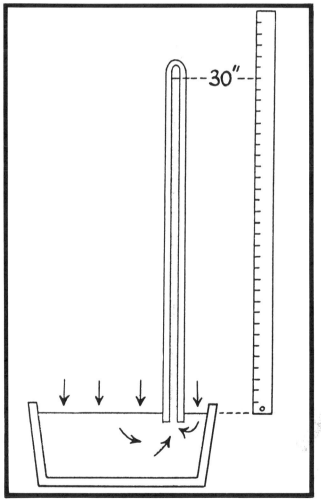

Figure 7

He reasoned that the force which was holding the mercury up was the weight of the air pressing down on the exposed mercury in the dish. It followed from this that the weight of the air pressing down on the mercury in the bowl was exactly equal to the weight of a column of mercury 30 inches long. It was then easy enough to determine that a column of mercury having a cross sectional area of 1 square inch and a height of 30 inches weighed 14.7 pounds. We know, therefore, that the air above us exerts a pressure of 14.7 pounds per square inch.

This device, invented by Toricelli, is still used for precise measurements of air pressure since air pressure is seldom exactly 14.7 pounds, but varies slightly, depending upon temperature, altitude and relative humidity. When this device is used to determine weather conditions, it is called a barometer; a modified form is sometimes used in the laboratory for precise measurements of low gas pressure, and in this form it is called a manometer. Both the barometer and the manometer indicate pressure by the height of the mercury column, and so they are directly calibrated to read in inches of mercury, rather than in pounds per square inch. However, it is a simple matter to convert inches of mercury into pounds per square inch if it becomes necessary to do so, as will be explained later.

GAS PRESSURE AND HOW IT VARIES WITH TEMPERATURE

As we know, a gas is composed of molecules darting around in every direction and at very high velocities. Although each molecule is almost unbelievably tiny, there are so very many of them in any measurable quantity of matter that the space between the molecules is extremely small. When a gas is contained in a vessel, these tiny, darting particles keep bumping into the walls of the vessel. Each time a molecule bumps into the wall, it imparts a tiny impact force. While the impact is very light and of extremely short duration, there are so many impacts due to the tremendous number of molecules that the result is a steady or constant push against the walls of the container. The sum of all these tiny pushes is the pressure exerted by the gas.

Earlier in this section we mentioned that the speed of the molecules in a gas is determined by their temperature. The higher the temperature of the gas, the faster is the movement of the molecules. More rapidly moving molecules will, of course, hit the walls of the containing vessel with greater impact, and this greater impact will, in

turn, exert more pressure. In other words, the pressure of a gas is proportional to its temperature.

This can be demonstrated very easily with the equipment shown in *Figure 8*. Here we have two containers, each fitted with a pressure gauge and a thermometer, and filled with any common refrigerant gas — Freon-12 will do very nicely. First the pressure and the temperature are checked at room temperature, see *Figure 8A*, after which heat is applied, see *Figure 8B*. It will be noted that as the temperature goes up, so does the pressure.

GAUGE PRESSURE AND ABSOLUTE PRESSURE

We use gauges to measure pressure. When the pressure we are measuring is higher than atmospheric pressure, we say the pressure is positive. When the pressure is less than atmospheric, we sometimes refer to it as a negative pressure, but more often we refer to it as a partial-pressure or a vacuum.

Ordinary pressure gauges are calibrated with zero representing atmospheric pressure. In other words, if the gauge is not connected to anything so that it is open to the air, it will read zero, even though it is actually reading atmospheric pressure. When the gauge reads 300 lbs., it is actually reading 300 lbs. above atmospheric pressure.

If it becomes necessary to make a distinction or to be more exact on this point, we use the term "gauge pressure" if we mean the pressure indicated by the gauge, and "absolute pressure" if we are starting from absolute zero. To convert gauge pressure to absolute pressure, we simply add 14.7 pounds.

In measuring vacuums, the gauge is calibrated in inches of mercury, rather than in pounds per square inch. This is because the original gauges which were used to measure pressures below atmospheric pressure were barometers or manometers, and used mercury columns rather than dials. These mercury columns were calibrated according to their length, rather than to the pressure they exerted.

Vacuum readings start from zero on the gauge, just as do pressure readings so that when a gauge reads, let us say, 30 inches of vacuum, it really means 30 inches below atmospheric pressure or approximately 15 pounds per square inch. (To

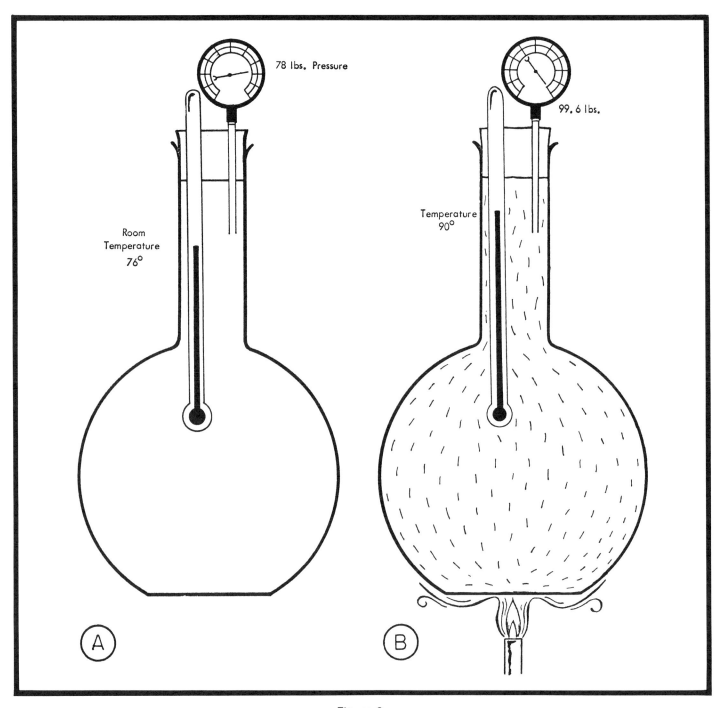

Figure 8

simplify our arithmetic, we will take atmospheric pressure to be 15 pounds per square inch instead of 14.7 pounds, since this is close enough for all practical purposes.)

If we think of the vacuum reading on the gauge being the same as a below-zero reading on a thermometer, we will have no trouble in understanding it. For example, 30 inches on the gauge is really minus 30 inches (−30 in.) or 30 inches below zero. It is, therefore, less pressure (a higher vacuum) than 20 inches.

1. To change gauge pressure (above zero) to absolute pressure, add 15.
2. To change gauge inches of vacuum (below zero) to absolute pressure, divide inches of vacuum by 2, then subtract answer from 15. Answer will be in absolute pressure.
3. To change absolute pressure (above 15) to gauge pressure, subtract 15.
4. To change absolute pressure (below 15) to inches of gauge vacuum, subtract absolute pressure from 15, then multiply answer by 2. Answer will be inches of vacuum.

For example, let us convert 20 inches of vacuum to absolute pressure. Twenty inches is equivalent to 10 pounds, so we subtract 10 from 15, and our answer is 5 pounds per square inch absolute.

In *Figure 9*, we show a gauge calibrated in both gauge pressure and absolute pressure so you will be able to understand them better. In *Figure 10* we show a conventional pressure gauge. Note that the zero point is all the way to the left so that this gauge cannot read any pressures below atmospheric pressure. In *Figure 11*, we show a gauge which can be used for both pressure and vacuum readings. When open to the air, this gauge will also read zero, but the zero point is some distance up the scale, so that the gauge can be used for pressures above or below atmospheric. In other words, this gauge will read both pressures and vacuums. Such a gauge is called a compound gauge.

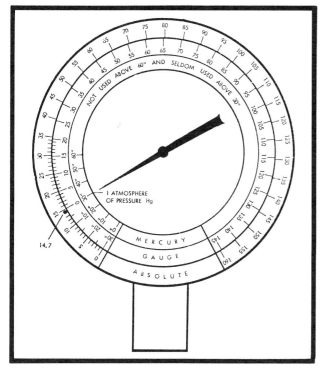

Figure 9

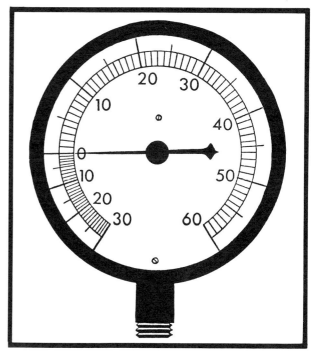

Figure 11

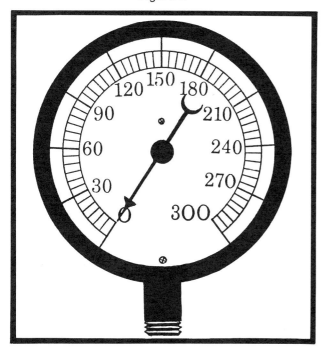

Figure 10

THE RELATIONSHIP BETWEEN PRESSURE, TEMPERATURE AND VOLUME OF A GAS

We are now ready to begin our imaginary experiment to show the relationship between the pressure, the temperature and the volume of a gas. For this experiment, we would need the equipment shown in *Figure 12*. This consists of a cylinder fitted with a gastight piston which can be raised or lowered as required. In the bottom of the cylinder are two openings, in one of which is inserted a pressure gauge. The second opening is connected to a drum of refrigerant, such as Freon-12.

Incidentally, since we are talking about refrigerants, we might give you a simple definition. A complete discussion of refrigerants will be given in a later section, but for the time being we can consider a refrigerant as a substance which is a gas at normal temperatures and which has a boiling point (boiling temperature) lower than the temperatures we wish to maintain in our refrigerating system. A very common refrigerant for use in domestic refrigerators is Freon-12, which has a boiling point in open air of −21.7°F.

To get back to the experimental set-up in *Figure 12* the refrigerant drum has a valve so that any

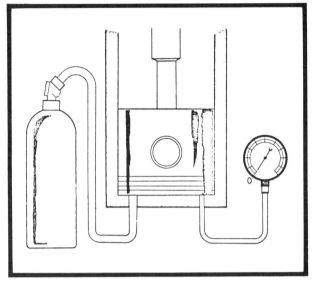

Figure 12

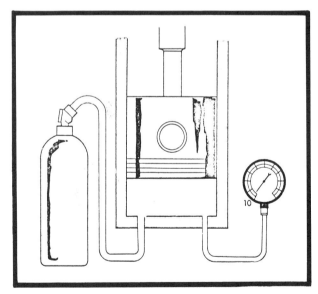

Figure 13

required amount of refrigerant can be added to the cylinder. We will begin our experiment with the cylinder completely empty so that the piston rests on the bottom. We now open the refrigerant drum to let some of the refrigerant enter the cylinder.

As the Freon enters the cylinder, it will cause the piston to rise. When the piston has risen a few inches, we will close off the valve on the

refrigerant drum and read the pressure gauge. Let us say the pressure is 10 pounds per square inch, as shown in *Figure 13* .

Now we raise the piston, thus increasing the volume occupied by the gas. You will note that the pressure decreases correspondingly; See *Figure 14*. Then, we let the piston drop back to its original position so that the pressure will again be 10 pounds. Now we exert pressure on top of the piston to force it down and reduce the volume. See *Figure 15*. If we look at the gauge, we will

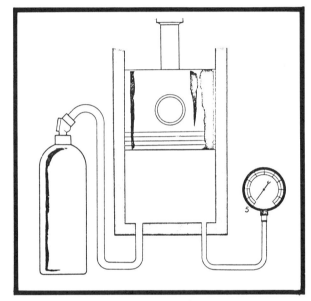

Figure 14

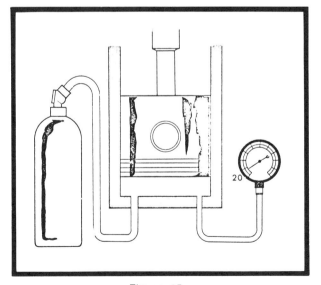

Figure 15

find that the gauge pressure has risen above the original value of 10 pounds per square inch. This demonstration shows that the volume of a gas varies inversely with the pressure.

Now let us repeat our experiment, but this time we will raise the piston very rapidly and watch the thermometer. We will find that as the volume is increased, the temperature drops; See *Figure 16*.

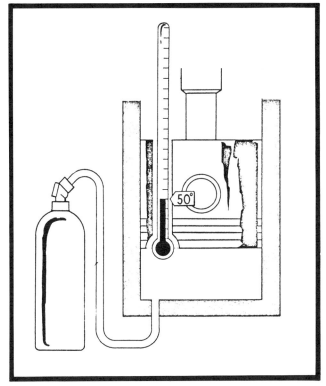

Figure 16

Now the volume is reduced suddenly by pushing down on the piston and compressing the gas rapidly. As the volume is reduced, the temperature rises, *Figure 17*. This teaches us that when a gas is permitted to expand, its temperature drops (it gets cold); when a gas is compressed, its temperature rises (it gets warm).

You will understand why this is so if you think back to our definition of temperature. Temperature is defined as a measure of heat intensity. If we let a gas expand so that it has a greater volume, the intensity of its heat will be reduced correspondingly, and its temperature will drop. On the other hand, if we compress it, we concentrate it so that the intensity of its heat is increased and the temperature rises. This is a basic principle which is used in all mechanical refrigeration systems.

FREEZING POINT OF SOLUTIONS

We know that pure water freezes at 32°F. What about impure water, or water which contains dissolved matter? You will recall that water becomes ice (freezes) when the movement of the molecules slows down sufficiently so that intermolecular attraction takes over and limits the freedom of movement to the point at which the liquid (water) becomes a solid (ice). This slowing down of the molecular movement is the result of lowered temperature. In the case of water, the specific temperature at which freezing occurs is 32°F. Other liquids freeze at other temperatures.

If a certain substance is dissolved in water, the molecules of that substance are uniformly distributed throughout the water and they move about in a random pattern in much the same way as the water molecules themselves. The movements of these foreign molecules, however, interfere with the normal behavior of the water molecules so that freezing is retarded. The temperature of the water must, therefore, be reduced before it will freeze.

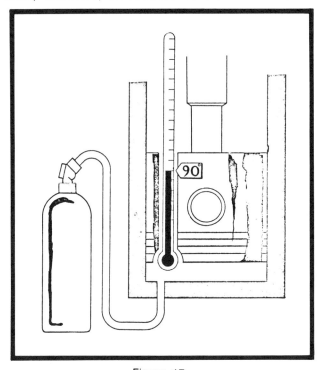

Figure 17

The lowering of the freezing point can be quite considerable if the solution is strong enough. For example, if we sprinkle salt on ice, the ice will melt unless the temperature is near zero, because the freezing point of a concentrated solution of salt in water is considerably below 32°F.

It is common practice to use antifreeze solutions in radiators of automobiles in winter. Such antifreeze preparations dissolve in the water of the car's cooling system and lower the freezing point by their presence. If the solution is concentrated, the freezing point of the water can be lowered to 20° below zero, or even lower without difficulty.

In certain types of refrigeration systems, it is sometimes necessary to circulate a coolant or cooling agent at a temperature slightly below the freezing point of water through the coils or in large tanks for the rapid cooling of liquids such as milk or beer. In the early systems, the coolant used was a strong solution of common salt in water, which is called brine. Although nowadays we find it better to use less corrosive solutions, such as those used in the radiators of cars, which are based on ethylene glycol or similar chemicals instead of salt, the term "brine" is still used as the name in the refrigeration industry.

BOILING POINT OF SOLUTIONS

Since the presence of dissolved matter in a solution lowers its freezing point, does it also lower its boiling point? Let us see if we can deduce the answer by visualizing the activity of the molecules in a solution.

As we know, the molecules of a liquid are darting about in all directions in a random pattern. From time to time, one molecule may collide with another in such a way that one of them is projected out of the liquid altogether. Evaporation is nothing more than the escaping of large numbers of molecules in just this way.

As the temperature of a liquid is raised, the activity of its molecules increases and, therefore, more and more of them escape from the liquid. A point is finally reached when the average velocity of the molecules is just about at the escape velocity. Any more heat added beyond this point causes more molecules to escape, rather than rise in temperature. This is the boiling point.

If there are foreign molecules dissolved in the liquid, they get in the way of the escaping molecules. The average velocity of the molecules of the liquid must, therefore, be increased to overcome this interference. Since temperature is an indication of the velocity of the molecules, this is just another way of saying that the temperature must be increased. Our basic principle will then be that the presence of dissolved material in a liquid raises the boiling point of that liquid. As an example, sea water, with its high percentage of dissolved salts, has a higher boiling point than fresh water.

BOILING POINT VS./ PRESSURE

When we are talking about the boiling point of water being 212°F., we are referring to pure water. The next question is, whether pure water always boils at 212°F., or whether it boils at different temperatures under certain conditions. Here, again, we can deduce our answer by visualizing the activity of the molecules.

Up to this point, when we thought of evaporation, we thought of it as the escaping of molecules from the liquid. We did not go any further and consider what happened to the molecules after they escaped. Actually, since there is such a vast quantity of molecules involved, there is always a large cloud of free molecules above any liquid. As a result, many of the escaping molecules collide with these free molecules and bounce back into the liquid. This, of course, reverses the process of evaporation to some extent. The greater the number of free molecules above the liquid, the greater is the number of escaping molecules which are bounced back. To offset this reverse process, the average velocity of the molecules within the liquid must be increased, and this means that the temperature of the liquid must be raised as well.

We cannot, of course, count the number of molecules above a liquid, but we do know that their presence is indicated by the pressure they exert. The greater the number of free molecules in any volume of vapor or gas, the greater is the pressure they exert. Since the temperature at which a liquid boils is raised in proportion to the number of free molecules above the liquid, it follows that the boiling point of a liquid is increased by an increase in pressure.

We can demonstrate this fact with the equipment shown in *Figure 18*. This consists of a container fitted with a thermometer and a pressure gauge which indicates both positive pressures and negative or partial pressures (such a gauge is called a compound gauge). The drum is fitted with a vent tube having a cross-sectional area of exactly one square inch. The vent can be closed off by weights.

We begin our experiment by pouring a little water into the drum and applying heat. As heat is applied, the temperature of the water steadily rises until it reaches 212°F., after which it levels off while the water starts to boil. If our container is made of a transparent material, we can see the boiling.

Figure 18

Figure 19

While the water is gently boiling, a one-pound weight is placed on top of the vent in order to close it. The following action takes place: The pressure within the drum builds up to one pound, at which point it will overcome the pressure of the one-pound weight on the vent, and steam will start escaping again. While the pressure is rais-ing, so is the temperature. When the pressure reaches one pound, the temperature reaches a value of approximately 215°F. The temperature will remain at this level as long as the pressure is one pound per square inch.

If we increase the weight to five pounds, the tem-perature of the boiling water will increase to around 227°F., *Figure 19* . In this way we can prove that for every value of pressure, there is a correspond-ing value of temperature at which a liquid boils. The reverse is also true — for a certain temperature there is one, and only one, pressure. This relation-ship is basic and cannot be changed.

Incidentally, the device we have described in the preceding experiment is nothing more than an ord-

inary pressure cooker. The thermometer is omitted, and sometimes spring pressure is used to shut off the vent instead of weights. In an ordinary sauce-pan, which is open to the air, the temperature of the cooking food is limited to slightly above 212°F. (The dissolved material in the water raises the boiling temperature slightly above 212°F.) In a pressure cooker, the temperature is raised to much higher values, greatly reducing the cooking time.

We must, at this point, bear in mind that when a gauge registers zero, there is actually a pressure of 14.7 pounds of atmospheric pressure. We must, therefore, ask ourselves the question whether the pressure-temperature relationship we have just demonstrated holds true for less than atmospheric pressure.

We can prove that it does with the set-up shown in *Figure 20* , which is, as you can see, substan-tially the same as in *Figure 18* , except that the vent is connected to a vacuum pump, and the de-gree of vacuum can be controlled by means of a valve in the vacuum or suction line.

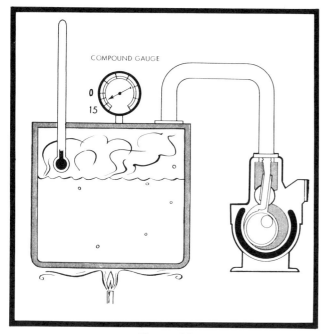

COMPOUND GAUGE

Figure 20

We adjust the size of our gas flame and the valve opening so that the pump draws away the steam at exactly the same rate as it is generated by the boiling water. Under these conditions, the gauge will read zero pressure and the thermometer will show a boiling point of 212°F. Now, let us open the vacuum pump valve so that the pump will draw away the steam at a faster rate than it is being generated. The pressure will drop. Let us say that the conditions are such that the pressure drops to 15 inches of mercury and levels off at that point, as shown in *Figure 20*.

As the pressure drops, the vigor of boiling increases, and since this excessively rapid boiling requires more heat than is being supplied by the flame, the temperature of the water drops even though the boiling continues. As the pressure levels off, so does the temperature, until at a pressure of 15 inches of mercury, the temperature stabilizes at about 179°F.

With this experimental set-up we can prove that for every value of pressure below atmospheric pressure, there is a corresponding value of temperature at which water boils. If we use a big enough vacuum pump, so that the pressure can be reduced enough, we can actually lower the boiling temperature to 32°F., the freezing point of water at atmospheric pressure.

There are, as a matter of fact, commercial air conditioning systems in which the refrigerant is

water, which is made to boil and absorb latent heat of vaporization at a temperature of only 40°F. by reducing the pressure sufficiently.

Although our experiment was done with water, the same basic principle holds true for all liquids. In other words, there is a basic relationship between pressure and boiling point for each liquid.

The refrigeration serviceman must thoroughly understand this relationship because it is by far his most useful tool in trouble-shooting a refrigeration or air conditioning system. Taking a pressure reading with a suitable gauge connected to that part of the system he is investigating can often tell him not only whether the system is working properly, but what is wrong with it. Because of the importance of this relationship, let us look at it from a slightly different point of view in order to get a more complete picture. We are referring particularly to substances which are gases at ordinary temperatures and pressures, but which can be liquefied with relative ease. Refrigerants come under this classification.

Let us take Freon-12 as an example. The boiling temperature of Freon-12 is −21.7°F. at atmospheric pressure (14.7 lbs. per square inch). When we buy a drum of Freon-12, it is in liquid form, and it is liquid because it is under pressure. Actually, the pressure in the drum is the pressure created by the gaseous Freon-12 above the liquid in the drum, and this pressure is always the pressure at which Freon-12 boils at the particular temperature the drum happens to be. Let us see how that works.

Suppose the drum is in a room at a temperature of 75°F. If we attach a pressure gauge to the drum, we will see that the pressure is 77 pounds. Now let us open the valve on the drum slightly in order to relieve the pressure to some extent. The Freon-12 gas will promptly start to escape, thus relieving some of the pressure. At this lower pressure, the liquid Freon-12 in the drum will start to boil, and in the process its temperature will start to drop because it is absorbing latent heat of vaporization, and the only way the latent heat can be obtained is by converting some of the sensible heat of the Freon-12 to latent heat. Since there is now less sensible heat, it follows that the temperature is lower.

If the valve is open only slightly, a new balance will be established between the pressure created by the boiling refrigerant and the release of that pressure through the open valve. Let us say that the valve is opened to a point at which this balance will be reached at a pressure of 12 pounds — in other words, the refrigerant will be boiling at a pressure of 12 pounds per square inch. The boiling point of Freon-12 at a pressure of 12 pounds is 6°F., and if we would take the temperature of the Freon-12, we would find it to be exactly 6°F.

If we close off the valve completely, the temperature of the drum will rise to room temperature and the Freon-12 will continue to evaporate until the pressure builds up again to 77 pounds.

If we heat the drum of Freon-12, the added heat will cause some of the liquid to evaporate, thus raising the pressure until this new, higher pressure corresponds to the new, higher temperature. On the other hand, if we cool the drum, we will reduce the pressure of the Freon-12 gas, because lowering the temperature of a gas lowers its pressure. At this lower pressure, there will be more gaseous Freon-12 than there should be, so some of it will condense and become liquid. In this way, if we raise or lower the pressure, the temperature automatically responds; if we raise or lower the temperature, the pressure automatically responds to maintain the basic temperature-pressure relationship.

THE REFRIGERATION SYSTEM

The earliest refrigerator for home use consisted of a box or cabinet in which was placed a large block of ice. As the ice melted, it absorbed heat of fusion from its surroundings, which were thus cooled. This was not only a clumsy way of obtaining refrigeration, but the minimum temperature obtainable was above that of melting ice. Actually, refrigerators of this kind, which were called ice-boxes, would seldom provide a temperature in the food compartment much below 50°F.

Let us see if we can develop a less clumsy system of refrigeration, and one which will permit much lower temperatures. Suppose, instead of the block of ice, we use a drum of refrigerant, such as Freon-12, which has a boiling point of −21.7°F. at atmospheric pressure. As shown in *Figure 21*, we attach a vent tube to the refrigerant drum and run the tube to the outside air. We then open the valve and close the cabinet. Since the boiling point of Freon-12 is much below the temperature of the interior of the box, the Freon-12 will im-

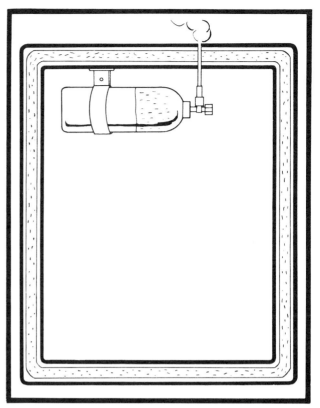

Figure 21

mediately start to boil. In the process of boiling, it will absorb heat of vaporization and get quite cold. If the vent is large enough, the pressure inside the drum will drop to zero gauge reading and the temperature of the liquid Freon-12 will drop to −21.7°F. We see, therefore, that this gives us a means of getting much lower temperatures than by using ice.

This device has two serious drawbacks. In the first place, there is no control of the temperature, and in the second place, Freon-12 is quite expensive, and letting it escape is highly wasteful. We can achieve a fairly good control of the temperature with the arrangement shown in *Figure 22*. Here we have put the drum of refrigerant outside the box and connected it to a coil of tubing through a fine needle valve, both the valve and the tubing being placed inside the box.

When we open the needle valve, we permit some Freon-12 to enter the coil of tubing. As it enters, it boils, or expands from a liquid to a gas. That is why this type of valve is called an expansion valve. As the refrigerant boils away in the tubing, it absorbs heat of vaporization, thus lowering the temperature of the tubing, which in turn lowers the temperature within the box. The tubing can, rightly, be called an evaporator because the refriger-

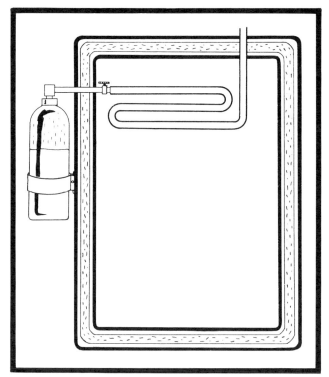

Figure 22

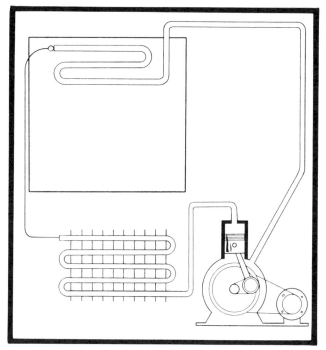

Figure 23

ant evaporates within it.

With this device we can control the temperature within the evaporator quite closely merely by regulating the valve opening. If the valve is wide open, the evaporator temperature will come quite close to the boiling point of Freon-12, which is −21.7°F. at zero gauge pressure. On the other hand, if the valve is shut off completely, the evaporator will warm up to room temperature. By regulating the valve opening, it is possible to attain any temperature in between these two extremes. We can, if we wish, work out some sort of automatic device which will open and close the valve as needed to maintain any desired temperature inside the box.

The next problem we must solve is how to save the Freon-12 so that it can be used over and again. How we can do so is shown in *Figure 23* .

The evaporator coil inside the cabinet, instead of being vented to the outside air, is connected to the intake of the pump. The outlet of the pump connects to a length of tubing arranged as a condenser or radiator, and the condenser, in turn, connects to a length of fine-diameter tubing, known as capillary tubing. The capillary tubing, because of its small diameter, restricts the flow of refrigerant in much the same way as does the needle valve in *Figure 22* . The other end of the capillary

tubing connects to the evaporator inside the refrigerator cabinet.

Those of you who may have had some experience with refrigeration servicing will recognize this diagram as a typical refrigeration cycle chart. If this is a typical refrigeration system such as is used in a domestic refrigerator, the system will contain about 1 lb. of Freon-12.

When this refrigerator is delivered to the customer, it is at room temperature. Let us say that this is 75°F., which means that the pressure in the system is 77 lbs. per square inch. The liquid Freon-12 could be anywhere in the system, depending upon how the refrigerator was handled before it reached the customer.

The refrigerator is now connected and the switch turned on. This starts the pump, and since the intake of the pump is connected to the evaporator, it begins to pump out some of the refrigerant gas from the evaporator. As refrigerant gas is removed, the pressure within the evaporator drops. If there is any refrigerant liquid in the evaporator, this liquid will immediately start to boil because of the lowered pressure. If there happens to be nothing but refrigerant gas in the evaporator, then liquid will shortly flow into it through the capillary tubing. In any event, the Freon-12 liquid will boil in the evaporator, absorbing heat of vaporization in the process, so that the evaporator will get cold.

The cold evaporator will, in turn, absorb heat from the warmer food in the refrigerator food compartment.

The Freon-12 pumped out of the evaporator is forced into the condenser tubing, where it is compressed into a high-pressure gas because the restriction presented by the capillary tubing prevents it from flowing through as rapidly as the pump can pump it out of the evaporator. Incidentally, because the pump in a refrigeration system is always used to compress the refrigerant in this way, it is usually referred to as a compressor, even though its function of lowering the pressure in the evaporator is really what starts the process of refrigeration.

As the compressor continues to run, more and more Freon-12 is pumped into the condenser, so that it becomes highly compressed and its pressure builds up. Earlier in this section we stated that compressing a gas concentrates its heat and raises its temperature. It is important for us to realize that this rise in temperature is not due to any heat being added to the refrigerant, but is due solely to the fact that the heat it contains is being concentrated. The hot refrigerant gas, at this point, contains the heat it started out with at the beginning of the cycle of operation, plus the heat it absorbed from the food compartment, but no more.

This highly compressed gas can be quite hot. In a typical refrigerator, the gas leaving the compressor may be above 200°F. As this hot gas flows through the condenser, it begins to cool off by giving up some of its heat to the surrounding air. At this point, the refrigerant actually contains less heat than it did when it entered the compressor, even though its actual temperature is very much higher.

Another thing that happens is that at this lower temperature, some of the gas condenses to a liquid and gives off the heat of vaporization which it absorbed when it boiled in the evaporator. By the time the Freon-12 has traveled through the condenser, all of it has become liquid and given off all the heat of vaporization it absorbed from the food compartment.

If we stop to think for a moment, we will see that what has happened here is that we have simply used the refrigerant to remove heat from the food compartment of the refrigerator and then we forced

the refrigerant to give up that heat in the condenser.

After condensation, the liquid Freon-12 continues to flow through the capillary tubing into the evaporator, where it boils again. This cycle continues as long as the compressor keeps running.

In order for this refrigeration system to work as it should, some restriction to the flow of the refrigerant must be placed at the entrance to the evaporator. This restriction prevents the refrigerant from flowing into the evaporator fast enough, so that it will maintain a certain low pressure in the evaporator as the compressor pumps out the gas from the other end. At the same time, the restriction causes the refrigerant to pile up in the condenser and capillary tube so as to raise its temperature and pressure, thus permitting it to give off its heat of vaporization.

Because the refrigerant from the outlet of the capillary tubing clear to the inlet valve of the compressor is at a low pressure, this part of the system is called the low-side, or suction side. The refrigerant from the output of the compressor through the condenser and all the way through the capillary tube is at a high pressure. This part of the system is, therefore, called the high-side. The capillary tubing offers the restriction necessary to

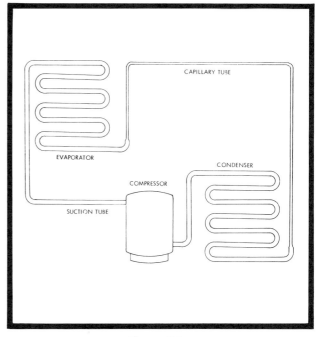

Figure 24

separate the high-side from the low-side outside the compressor. Capillary tubing is used almost universally for this purpose in modern domestic refrigerators and window-type air conditioners, but in larger commercial systems, an adjustable valve, called an expansion valve, is preferred. *Figure 24* is the cycle chart of a typical conventional refrigerator.

There is, of course, far more to a refrigeration system than we have discussed here. The various components will be discussed in greater detail in succeeding sections. In this section we have presented a brief rundown of how these components work so that you will get a better understanding of the refrigeration cycle and why a refrigerator gets cold.

MASTERING THE TRADE

The mark of a good Serviceman is not necessarily shown by the number of tools he has in his tool kit, but rather by the care he gives his tools and the manner in which he uses them. The proper use of tools not only identifies a good serviceman, but it also indicates a natural desire to do good work. This in turn makes it possible to do the work in the shortest time and in the best manner.

It is also very important to use the correct tool for a particular operation, because it is often the misuse of a tool, or using the improper tool, that ruins or damages the equipment, or results in an accident. A screw driver, for example, with a poorly shaped blade will slip; a badly damaged wrench, or a wrench of improper size, will do the same. Tools that are not properly cared for and kept in condition are real causes of accidents.

There are a number of special tools used in refrigeration and air conditioning work in addition to many common ones. There is also special test equipment which is not common to many other mechanical or electrical trades.

COMMON TOOLS

Let us first consider those tools which are common to most trades, such as wrenches, hammers, screw drivers, and files. There is a large number of wrenches on the market, and each type of wrench is designed for a specific purpose. A four-point box or socket wrench, for example, would be the best kind to use on a bolt with a square head because there would be turning pressure applied to all four sides rather than just to the corners, as would be the case if a twelve-point wrench were used.

All cap screws and bolts with hexagonal heads as hexagonal nuts should be loosened and tightened with a six-or twelve-point box wrench for the sake of safety. Twelve-point wrenches are more popular because they can be used in tight places where only one-twelfth of a turn can be made at a time. An open-end wrench is used in places where box wrench cannot be used. Such a place would be on a fitting with which a length of tubing is secured. Typical box wrenches are shown in *Figure 25* .

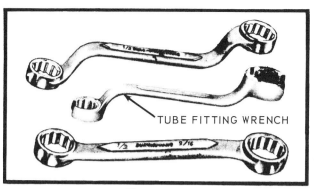

TUBE FITTING WRENCH

Figure 25

The adjustable wrench is well suited as a utility wrench because it eliminates the need of carrying box and open-end wrenches of every standard size in the tool chest.

When an adjustable wrench is being used, we must remember to place the wrench on the nut in such a manner that the pulling force is applied to the stationary jaw, as shown in *Figure 26* . After the wrench has been properly placed on the nut, the wrench must be tightened so that it grips the nut snugly.

When tightening nuts bolts, or cap screws, it is advisable to use a standard box wrench if posisble. The length of the standard wrench is such that the nut or bolt can be tightened sufficiently without undue strain on the threads of the nut or the bolt. As you tighten a nut or bolt you can tell by the "feel" when it is tight enough. Any further tightening will only damage the threads on the bolt or in the nut. The same thing applies when making flare connections. As you work with your wrenches, you will become accustomed to the "feel" when a bolt is tight enough.

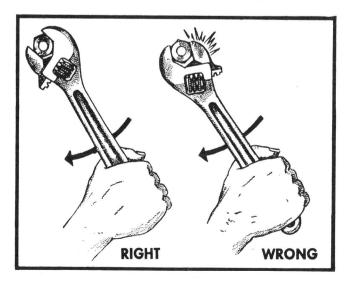

RIGHT   **WRONG**

Figure 26

Do not use a wrench of any kind for a hammer. It would damage the wrench, because wrenches are not made for that purpose.

One of the most common tools is the hammer. There are many kinds of hammers made for special purposes. The refrigeration serviceman has use for two: an all purpose hammer, such as a ball pein; a soft face hammer, such as plastic or lead, for use on brass parts or machined surfaces and screw threads. A ball pein hammer and a plastic head hammer are shown in *Figure 27*.

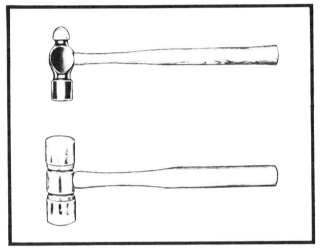

Figure 27

The most abused tool in a serviceman's kit is usually the screw driver. When a screw driver is used as a chisel or a prybar it destroys the sharp corners of the blade. This makes it difficult to hold the blade properly in the head of a screw. The blade of a screw driver will wear and is, therefore, in need of attention from time to time in order to keep it in good condition. When a screw driver is reground, it should be ground to a long taper, or, preferably, it should be slightly hollowground so that the two sides of the blade are almost paralled near the tip instead of being tapered.' An improperly ground screw driver will often slip out of the screw slot and damage the surrounding surface as well as the head of the screw.

The file is another necessary tool in the serviceman's kit. Files are manufactured in a variety of shapes and sizes and are meant to be used only on untempered metals. If a file is used on hard steel, for example, the teeth become dull within a short time, and the file is ruined.

When a file is used to cut very soft metal, such as

copper, it soon becomes clogged and loses its effectiveness. The file can be easily cleaned again, however, by flattening the end of a short piece of copper tubing and using it to rub across the file's teeth in line with the teeth. This will soon cut small ridges in the copper that will go down into the hollows between the teeth in the file and clean out any foreign substances. A file brush also can be used to clean a file. An occasional cleaning of the file is advisable to keep it in condition.

Special Tools

Among the special tools, probably the most valuable one to a serviceman is the "valve stem ratchet wrench", because this wrench is made especially for opening and closing the service valves on a compressor.

There are several designs of valve stem ratchet wrenches, two of which are shown in *Figure 28*. A valve stem should be turned only with a ratchet wrench, or with a socket wrench that fits the valve stem perfectly, because once the corners of the valve stem become rounded, it is necessary to cut the stem down to the next smaller size, or else replace the stem in the valve.

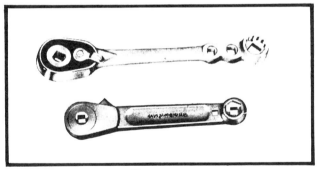

Figure 28

The handle of a valve stem ratchet wrench is usually provided with square openings of different sizes that will fit the most common valve stems used on refrigerant cylinders and on service valves. These square openings are used when a close control of the valve is needed during servicing, so that the valve may be quickly opened or closed as required without having to move the ratchet every time the turning motion is reversed.

The pawl and the ratchet wheel should be oiled occasionally in order to keep the wrench in perfect order and to insure long service.

A ratchet wrench should never be used as a hammer or prying device, nor should an extension be used on the handle in order to increase the leverage. The ratchet device will only withstand the the pressure received when the operator applies moderate pressure to the end of the handle itself.

The "tube cutter" is another special tool. This tool is used for cutting soft-drawn copper tubing. See *Figure 29*. A tube cutter is very similar to the plumbers pipe cutter, except that it is designed to cut only copper tubing and is not heavy enough for cutting harder metals. The cutting wheel is made of a special, highly tempered steel and is ground to a sharp edge. Therefore, the wheel should not be forced sideways during the cutting operation.

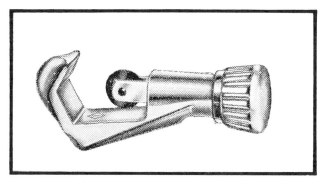

Figure 29

Some tube cutters are equipped with a retractable reamer which serves to remove any burrs left on the inside of the tube during the cutting operation, See *Figure 30* . The reamer also is made of tempered steel so that it will hold an edge and it must be handled just as carefully as the cutting wheel in order to prevent the point of the reamer from breaking off.

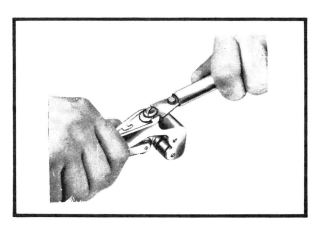

Figure 30

Many tube cutters are also equipped with rollers which serve to reduce the friction between the tubing and the cutter itself as the cutter is being turned while the cutting wheel exerts pressure against the tubing. These rollers are often provided with a special flare cut-off groove, as shown in *Figure 31* , so that a damaged flare may be cut off with a minimum amount of waste.

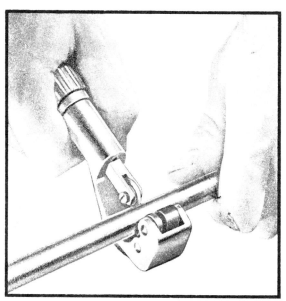

Figure 31

The proper method of cutting copper tubing is to set the cutting wheel against the tubing with very little pressure and rotate the cutter around the tubing and at the same time tighten the wheel adjustment slightly during each revolution, as shown in *Figure 32*. If too much pressure is exerted on the adjusting handle, it will have a tendency to crush the tube, or cause an excessive burr on the inside of the tube.

Figure 32

Heavy walled tubing and tubing of very large diameter, as well as steel tubing, should be cut with a hacksaw in conjunction with a special sawing vise, as shown in *Figure 33*.

Figure 33

The "flaring tool" may also be considered as a special tool. This tool is used in making the cone-shaped flare on the ends of the tubing that is used for connections. This flare serves as a soft copper gasket between two flare fittings and provides a gastight joint when these fittings are drawn together. A flaring tool consists of two parts, namely, the yoke and the block, as shown in *Figure 34* . The block, in turn, consists of two halves. These two halves are matched so that one half fits against the other in perfect alignment. When the two halves are properly aligned, they form holes of different diameters corresponding to the standard sizes of tubing used in the refrigeration and air conditioning industry. The top part of each hole is countersunk so that the tubing may be flared to the exact angle of the seat in the fitting. See *Figure 35* .

The yoke is made to slide freely over the flaring block in order that the cone at the end of the adjusting screw may be lined up with any one of the holes in the block.

In making a flare with this type of tool, the tubing is placed in the corresponding size hole in the flaring block and with the end protruding slightly above the face of the block. Care must be taken so that the tubing is clamped tightly in the block in order to prevent the tubing from slipping when

the cone is screwed down. Slipping of the tubing would result in an incomplete flare. The flare is produced by placing the cone over the tubing and turning the cone down until the flared part of the tube fits solidly against the countersunk recess in the flaring block.

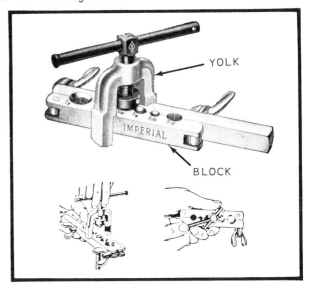

Figure 34

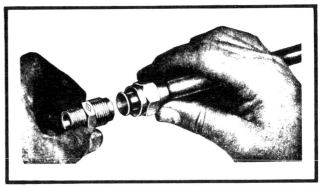

Figure 35

The cone of a new flaring tool should be lubricated with a light coating of oil until the cone becomes highly polished and able to slide smoothly without further lubrication. (In some yokes, the flaring cone does not turn when the cone is screwed down. These cones need not be lubricated.) The thread on the yoke should also receive an occasional oiling in order to reduce wear and to prevent stripped threads. At no time should the cone be screwed down tightly against any one of the holes on the reverse side of the flaring block where the holes are not countersunk. This would cause a ring to be cut in the cone and would make it impossible to produce highly polished flares. This flaring block should never be used as a clamping device for anything except copper tubing, nor should it be hammered on, or used as a hammer. If the countersunk recesses on the surface of the flaring block become only slightly damaged, an imperfect flare will result which will inevitably lead to a leak even if the connection is drawn tight.

An important part of a serviceman's job is to make sure that all connections are perfect connections and in perfect condition. We cannot stress too strongly the importance of making the flare joints properly. One improper flare connection may cause an entire system to fail. The joints may not be so bad that it will be apparent through a casual inspection, but a more thorough inspection may reveal weaknesses in a certain joint or joints. Each connection should be checked very carefully as soon as it is made, and if it is a new system, the entire system should be checked and pressure tested before it is put into operation. Considerable practice is needed before proper flare connections can be made.

There is still another tool which is used almost exclusively in the refrigeration and air conditioning servicing business, namely, the "bending spring". A bending spring is nothing more than a long coil spring wound to the proper diameter, so that it will slide over the tubing. For tubing of larger diameter, there are bending springs provided that will slide inside the tubing instead of over it. Those bending springs which are meant to slide over the tubing have the last few coils at one end of the spring shaped into a funnel to make it easier to insert to remove the tubing, whereas, the inside types are uniform throughout. *See Figure 36*.

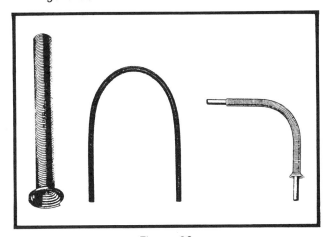

Figure 36

The purpose of the bending spring is to prevent the buting from kinking when making a short bend.

At some time or other, you have no doubt tried to bend a thin-walled piece of tubing and found that it would crimp or kink. This is because the metal in the outside part of the tube at the bend must stretch while the metal in the inside part must compress in order to form the bend. The bending spring prevents the outside of the tubing at the bend from spreading, and prevents the inside part of the bend from bucking.

When making a bend with an external bending spring, the spring is slid over the tubing and centered over the approximate spot where the bend is to be made. The spring and tubing are then gripped firmly with both hands and with the thumbs meeting each other in the bend. The bend is formed gradually as shown in *Figure 37*. It is always necessary to bend the tubing a little more than is needed and then straightening the tubing slightly tends to releave the pressure of the tubing against the spring. This in turn makes it easier to remove the spring.

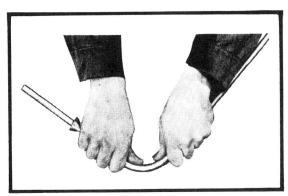

Figure 37

There are also a number of mechanical benders on the market. A very popular "lever-type" bender as shown in *Figure 38*.

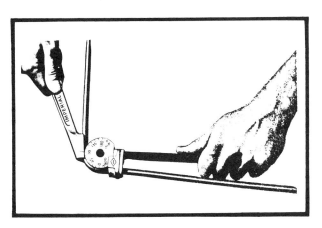

Figure 38

A mechanical bender is especially useful when tubing is to be bent to a short radius and to accurate dimensions. The mechanical bending tool does a much neater job and is, of course, far superior to the bending spring when a large number of bends is required. *Figure 39* shows another type of mechanical bending tool which is provided with several interchangeable wheels that will bend tubing of various sizes. Small-size tubing, up to and including 1/4" in diameter, can, as a rule, be bent by forming the bend over the two thumbs.

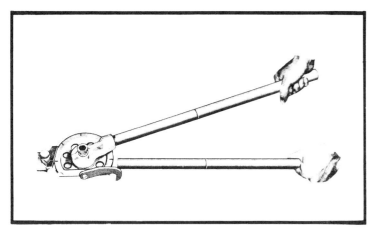

Figure 39

When bending tubing 3/8" to 1/2" diameter, a bending spring should always be used. Bending tubing larger than 1/2" by the use of a bending spring is not recommended. In such cases, the mechanical tube bender is used.

FLARE FITTINGS

Most flare fittings are made of brass. The fittings are first drop-forged to close dimensions. The threading is then done by machine to the required tolerances in order to assure gas-tight connections.

Certain types of flares fittings, particulary the flare nut, may also be made of cadmium-plated steel, but brass is still the most satisfactory alloy.

The various flare fittings have special names, depending on the application. For this reason, it is very important that you learn to recognize such a fitting on sight so that you can tell at a glance both the size of the fitting and where it belongs. *See Figure 40*.

Flare fittings can be divided into two main groups,

namely, male and female fittings. A flare fitting which has the threads on the outside is called a male fitting, whereas a fitting with the threads on the inside is referred to as a female fitting.

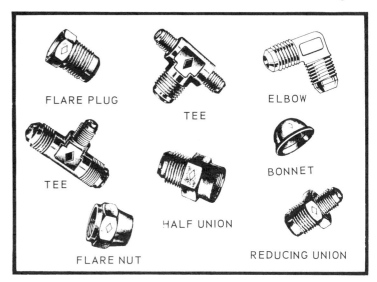

Figure 40

The standard flare fittings, such as flare nuts, flare caps, flare union, etc., are never referred to as "male" and "female" fittings. Instead, they are identified by their adopted trade names. A flare union, for example, is in reality a "male flare by male flare connector" but it is always referred to as a "flare union." Also, the connector that is used to connect a flare copper tubing to a threaded iron pipe is called a "half union" rather than a "male flared by male iron pipe connector."

The purpose of using standard names and descriptions for the various fittings is, of course, to make it possible for everyone connected with the industry to understand exactly what fitting or fittings are being referred to when an order for fittings is placed, for example, or when the specifications for an installation are being consulted.

Those fittings which we refer to as unions and half-unions may be classified into five main groups, if we consider the type of thread used and whether the thread is on the inside or on the outside.

1. The fitting may have male flare threads at both ends.

2. The fitting may have male pipe thread at one and male flare thread at the other.

3. The fitting may have female pipe thread at one end and male flare thread at the other.

4. The fitting may have female flare thread at one end and male flare thread at the other.

5. The fitting may have male pipe thread at one end and female flare thread at the other.

One important thing to remember is that pipe thread and flare thread are entirely different, both as far as the number of threads per inch is concerned, as well as the diameter of the threaded part itself, although there are certain fittings in both categories that approach each other closely enough in diameter and threads per inch to confuse a beginner. Always be very careful when making a connection to be sure you are using two fittings that are designed to be together.

When the plumbers and steam fitters established standards for pipe sizes, they did not go by the exact, measured diameter. There is, therefore, a great discrepancy between the so called nominal diameter and the actual diameter, as shown in *Figure 41*. An iron pipe with a nominal diameter of 1/2 inch, for example, will have an actual outside diameter of almost 1 inch.

I.P.T. is the abbreviation used in the plumbing industry for "iron pipe thread," and I.P.S. means "iron pipe size". F.P.T. stands for "female pipe thread." The abbreviations "F.I.P." and "M.I.P." are used to identify pipe fittings. F.I.P. means that the fitting has a female iron pipe thread in the fitting. M.I.P. means that the fitting has male iron pipe threads on the end of the fitting.

The next flare fitting we are going to discuss is the "Tee." Since a flare Tee has three outlets, as shown in *Figure 40*, it is necessary that we use a uniform system of identifying these outlets. The two outlets that are feeding straight through the Tee are known as the "Run," and the outlet which forms right angles with the other two outlets is referred to as the "Branch." For example if we were ordering a flare Tee for three different sizes of tubing, and we wanted to connect the largest size tubing to one end of the run, the next largest size to the branch, and the smallest size to the other end of the run, we would call out the two sizes on the run first, and the size on the branch last. Let us say that the three dimensions are 5/8 inches, 3/8 inches and 1/2 inch. We would then identify the Tee as "5/8 inches by 3/8 inches on the run plus 1/2 inch on the branch." Notice

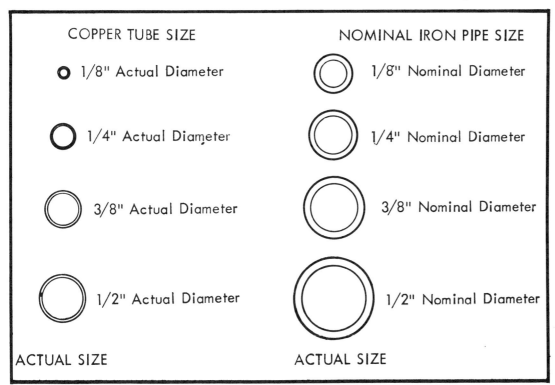

COPPER TUBE SIZE

O 1/8" Actual Diameter

1/4" Actual Diameter

3/8" Actual Diameter

1/2" Actual Diameter

ACTUAL SIZE

NOMINAL IRON PIPE SIZE

1/8" Nominal Diameter

1/4" Nominal Diameter

3/8" Nominal Diameter

1/2" Nominal Diameter

ACTUAL SIZE

Figure 41

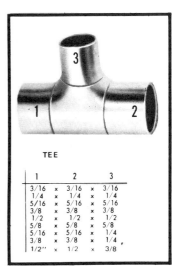

TEE

1		2		3
3/16	x	3/16	x	3/16
1/4	x	1/4	x	1/4
5/16	x	5/16	x	5/16
3/8	x	3/8	x	3/8
1/2	x	1/2	x	1/2
5/8	x	5/8	x	5/8
5/16	x	5/16	x	1/4
3/8	x	3/8	x	1/4
1/2"	x	1/2	x	3/8

Figure 42

that the branch dimension is always given last. This means of identification is used for all Tees regardless of size, *Figure 42*.

The sizes of standard flare fittings range from 3/16 of an inch to 3/4 of an inch. These sizes are: 3/16 in., 1/4 in., 5/16 in., 3/8 in., 7/16 in., 1/2 in., 5/8 in., and 3/4 in. The 5/16 in. and 7/16 in. sizes are not in common use at present, but they were used quite extensively a few years back by several manufacturers. The actual diameter across the thread, as well as the number of threads per inch for these flare fittings, are illustrated in *Figure 43*.

FITTING SIZE	DIAMETER ACROSS THREADS	THREADS PER INCH
3/16"	3/8 "	24
1/4 "	7/16"	20
3/8 "	5/8 "	18
1/2 "	3/4 "	16
5/8 "	7/8 "	14

Figure 43

SWEAT OR SOLDER FITTINGS

There is another kind of fitting which is used quite extensively in the industry, namely, the sweat or solder fitting. *See Figure 44*. Sweat or solder fittings are used primarily on larger units and where the connections are exposed to a great amount of vibration and strain. Tests have proved that the strength of a soldered connection is as strong as, or stronger than, the lengths of tubing being joined by the fitting in withstanding pressure and continuous vibration. Most of these fittings are now made of wrought copper, whereas the earlier ones were made of drop-forged brass, and machined to fit the tubing.

Sweat or solder fittings are made in all the standard sizes mentioned for flare fittings, and there is just as great a variety of sweat or solder fittings as there is of flare fittings. We have adapters, Tees, and elbows with different degrees of bend, etc. We also use the same system of identifying a sweat or solder Tee, for example, as we do for a flare Tee, *Figure 42*.

In *Figure 44* are shown two reducing Tees; one has the reduction on the branch and the other on the run. The first-mentioned Tee would be referred to as a "sweat or solder Tee, 1/2 in. by 1/2 in. on the run plus 3/8 in. on the branch," and the other as a "sweat or solder Tee, 1/2 in. by 3/8 in. on the run plus 1/2 in. on the branch.

In most cases, a sweat or solder fitting is made to slide over the ends of a tubing instead of having the tubing slide over the fitting.

TUBING

Soft-drawn copper tubing and hard-drawn copper tubing are both used in refrigeration and air con-

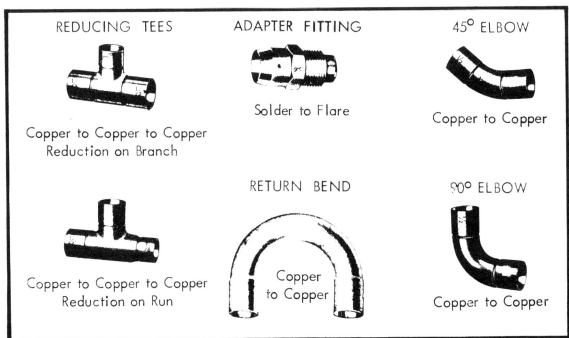

REDUCING TEES

Copper to Copper to Copper
Reduction on Branch

Copper to Copper to Copper
Reduction on Run

ADAPTER FITTING

Solder to Flare

RETURN BEND

Copper to Copper

45° ELBOW

Copper to Copper

90° ELBOW

Copper to Copper

Figure 44

ditioning installations. The soft drawn tubing is usually sold in rolls, as shown in *Figure 45*, and with both ends of the roll sealed by being crimped together. The reason the ends are sealed is because the tubing has been heated in order to remove all moisture and sealing the ends assures that the tubing will be perfectly clean and dry when it is ready for use.

Figure 45

When a piece of tubing is to be cut from a roll, the proper procedure is to first crimp the tubing with a hammer a short distance beyond the required length. The tubing is then cut with a regular tube cutter. This will prevent any moisture or foreign substances from entering the remaining roll.

Soft-drawn tubing can be bent and shaped very easily. A neat bend that will offer little resistance to the flow of refrigeration can be shaped very readily with the aid of a bending tool or a bending spring. It might be well to mention in this connection that when a piece of tubing has been bent and then straightened a number of times, it gets hard and crystalline. This is one reason why you should try to make the proper bend on the first attempt, because it becomes increasingly harder to form the bend after each time the tubing has been straightened.

Hard-drawn copper tubing cannot be readily formed into sharp bends, except with a tube bender. Most bends in an installation using hard-drawn tubing are made with sweat or solder fittings, except for slight offsets that can be made without any risk of flattening the tubing.

Hard-drawn tubing is usually sold in 20—foot lengths. Some manufacturers seal the ends of the tubing with plastic caps, or by some other

means, whereas other manufacturers pay very little attention to keeping the tubing dry and clean on the inside.

Plastic tubing is also used to a limited extent in the refrigeration and air conditioning industry. This type of tubing is better known by the various trade names assigned by the manufacturers.

Plastic tubing is sometimes used instead of copper tubing when more flexibility is needed, or when it is desired to see the flow of the refrigerant within the system. Plastic tubing is transparent so that the rate of flow of the refrigerant or oil can be observed while the unit is running.

Flaring of plastic tubing is done in the same manner as copper tubing. However, plastic tubing tends to become brittle at lower temperatures. It burns in the presence of a flame, and is not easily flared at temperatures below 100° F.

The method to use in flaring plastic tubing is to warm the flaring tool before making the flare. Since plastic tubing will burn, the flaring block must be heated with a blown torch or other open flame after the tubing has been inserted into the block. Flare fittings of molded plastic are sometimes used, although brass fittings can be substituted if reasonable care is excercised in tightening the fittings on the plastic tubing.

Flaring Copper Tubing

An important part of servicing consists of replacing damaged or leaking refrigeration lines in older system, as well as making new ones in installations. This, in turn, involves fabrication connections of various kinds that will remain gas-tight and secure for long periods of time. We will begin this discussion with flare connections and how these connections are made.

Before beginning the actual flaring operation, you must make sure that the end of the tubing is approximately round and that it has been cut off squarely. Then place a flare nut of the proper size on the tubing and slide the nut away from the end of the tubing with the threaded end of the nut facing toward the end to be flared. Next, place the end of the tubing in the flaring block while making sure that the hole in the flaring block and the tubing correspond in size. *See*

Figure 46 . Also make sure that the countersunk portion of the hole in the flaring block is facing toward the end of the tube to be flared. Then clamp the block over the tubing and tighten, as shown in *Figure 46* . The amount of tubing extending above the flare portion of the hole will determine the size of the flare. Only practice will tell you the correct amount, but in every case, the flare must at least cover the countersunk portion of the hole.

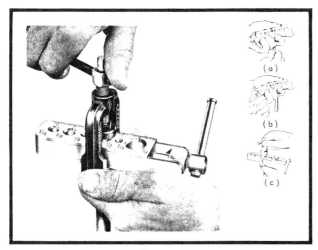

Figure 46

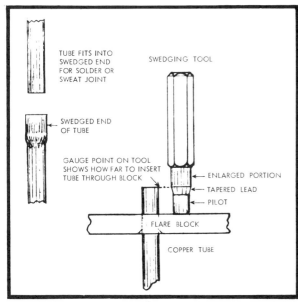

Figure 47

The next step is to slip the yoke over the flaring block, as shown in *Figure 46* . The cone of the yoke is now centered over the end of the tube and drawn snugly into the flare, as shown in the enlarged illustration of *Figure 46* . If the right amount of pressure has been used, the flare will have the correct angle. On the other hand, if too little pressure has been applied, the flare will have an arc instead of being angular.

After the flaring block has been removed, inspect the flare in order to make sure that it is free from cracks and rough spots. The inspection should also reveal whether the flare nut will slide over the flare without interference from the threads, and whether it will turn easily on the end of the tubing. We cannot over-emphasize the importance of making good flares because one bad flare may cause a complete system to fail.

The next step is to screw the flare nut onto the fitting. The nut should never have to be forced onto the fitting with a wrench. This will damage the threads, both in the nut and on the fitting. If the flare and the fitting are aligned properly from the start, you should be able to turn the nut by hand until it begins to tighten on the flare.

In tightening the connection, it is well to remember that the nut should be turned and not the fitting. The joint is produced by the flare forming a sealing gasket between the beveled end of the fitting and the shoulder of the nut. This is a different type of seal than is produced when a threaded pipe is screwed into a companion thread. Pipe thread is tapered and will get tighter as the pipe is screwed further into a companion thread.

Flare threads, on the other hand, are straight and the seal is produced by the pressure of the fittings against the flare itself, and not by the threads fitting snugly together.

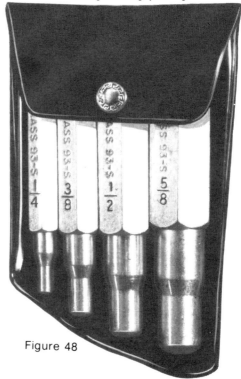

GEMLINE
TB-193-S

Figure 48

HOW TO USE THE SWAGING TOOLS

In making repairs to a refrigerator system it often becomes necessary to expand one tube to fit onto another for the purpose of brazing. A proper swage would allow the smaller diameter tube to enter the larger diameter tube at least the diameter of the tube. If the larger diameter tube does not go this deep there is a fair chance that vibration could cause the tubing to break at the point of braze.

For the purpose of swaging, the punch type swaging tools are used. Such a set of swaging tools is the Gemline, Part No. TB-193-S as shown in *Figure 48*.

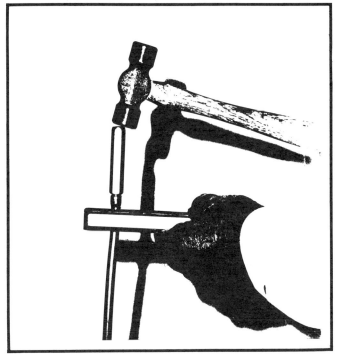

Figure 48 A

The tubing should be cut with a tubing cutter, *Figure 29* and the end properly reamed. Avoid deep nicks at the end of the tube as these will sometimes become deep cracks when swaging. The end of the tube should be smooth and the inside diameter of the tube should not be narrowed by the burr prior to swaging. Use the reamer on the tubing cutter to clear the tubing

Select the size swaging tool that you wish to expand the tubing. Place the tubing in the flaring tool, using the size hole that fits the tubing, allow the tubing to extend not more than ¾ of an inch above the flaring tool. Use the side of the flaring tool that is not countersunk. Insert the swaging tool into the tubing and strike with a hammer until the swaging tool enters up to the shoulder. To allow for the brazing material, slightly tap

the tool from side to side. Remove the tool and fit it to the smaller diameter tube. If the fit is not loose enough to allow brazing material, the tube will have to be expanded a little more, *Figure 48 A*.

Process Tube

Pinch-off Tool

Figure 48 B

SOLDERING

It requires just as careful a preparation when two lengths of tubing are to be joined by a sweat or solder fitting as when they are joined by a flare fitting. Here, again, the ends of the tubing must be approximately round and cut off squarely. The outside of the tubing is then cleaned with a piece of sandcloth or emery cloth. The inside of the fitting may be cleaned with steel wool or with a piece of emery cloth wrapped around a piece of wood. A small amount of soldering flame is then applied on the end of the tubing with a brush. The fitting is now placed over the end of the tubing and turned around several times in order to distribute the flux properly. The tubing should be inserted in the fitting until it reaches the shoulder. Do not touch any of the clean surfaces, or apply soldering flux with your hands.

The joint is then heated until the melting point of the solder is reached. Apply heat to both the fitting and the tubing with a wide flame, as shown in *Figure 49* Hold the torch in one hand and the wire solder in the other hand. Touch the solder to the fitting from time to time in order to determine when the fitting is hot enough to melt the solder. Remember, the solder will run to the hottest point. Therefore, it is best to heat all the parts of the fitting which have any contact with the tubing.

The joint is then heated until the melting point of the solder is reached. Apply heat to both the fitting and the tubing with a wide flame, as shown in *Figure 49*. Hold the torch in one hand and the wire solder in the other hand. Touch the solder to the fitting from time to time in order to determine when the fitting is hot enough to melt the solder. Remember, the solder will run to the hottest point. Therefore, it is best to heat all the parts of the fitting which have any contact with the tubing.

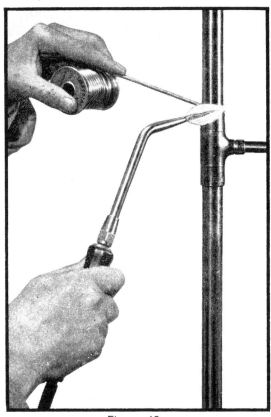

Figure 49

Care should be taken so that only sufficient heat is applied to the fitting to melt the solder. Overheating the fitting will cause it to oxidize and solder will not adhere to oxidized metal.

When the fitting is hot enough to melt the solder, continue to work the flame around the seam where the tubing enters the fitting and begin to apply solder. Solder is applied until it can be seen all around the seam. Finally, take a flux brush and brush over the top of the solder to give the joint a finished appearance.

The opposite end of the fitting is soldered in the same manner, even though it is sometimes in an inverted position in relation to the tubing. The solder will flow up into a joint just as readily as it will flow down into it. This upward flow of the solder is caused by capillary attraction due to the close contact between the tubing and fitting.

Solder is an alloy of different metals. The two most commonly used metals are tin and lead. Antimony is sometimes alloyed with tin and lead in order to change the hardness and the melting point of the solder. The ratio between the tin and the lead contained in a particular solder is expressed as a percentage. For example 40-60 solder contains 40% tin and 60% lead. Likewise, 50-50 solder contains 50% tin and 50% lead. 95-5 solder, on the other hand, contains 95% tin and 5% antimony.

Silver solder is used quite frequently because of its high tensile strength and ability to resist corrosion by acids and alkalies.

Silver solder produces a very hard joint, but it requires a high concentration of heat in order to melt it.

Since an ordinary gasoline blow torch or propane torch does not produce sufficient heat for this operation, it is necessary to use an Oxy-acetylene torch.

Brazing, or soldering with hard solder, is often used with hard-drawn tubing in order to produce a very rigid joint that is not affected by vibration or the elements. Brazing requires still hotter temperatures than silver soldering because it is done with brazing rods, such as copper alloyed with silver and phosphorous. Brazing must therefore be done with an Oxy-Acetylene torch.

WHAT IS SILVER BRAZING?

Silver brazing is defined as the joining of metals by means of heat wherein the filler metal is a silver alloy having a melting temperature above 1000°F. but below the melting point of the metals to be joined.

For many years silver brazing has been incorrectly called silver soldering. Soldering is the joining of metals by means of heat wherein the filler metal melts at a temperature below 900°F.

This confusion in terminology is undoubtedly the result of an old misconception that silver brazed joints more particularly resembled soft soldered joints as far as strength, ductility, corrosion resistance, etc., are concerned. On the contrary, as we shall see later, silver brazing actually falls in the category of high temperature brazing and welding.

WHAT WILL IT DO?

Silver brazing is probably the most versatile method of metal joining in use today. Silver brazed joints have high strength on non-ferrous metals and on mild and cast steel, the strength of a properly made joint will exceed that of the metals joined.

Silver brazed joints are ductile and as a result will withstand considerable vibration and shock.

Silver brazed joints are easy to make. By following six simple rules which will be covered later, it is possible for even an inexperienced person to develop speed and proficiency in its use in a comparatively short time.

SILVER BRAZE "45"

These are 45% silver alloys with a melting point of 1125°F. and flowing at 1145°F. These alloys not only have the advantage of a low melting point, but because they are Cadmium free the toxic effect of Cadmium fumes are eliminated.

SILVER-PHOSPHOROUS BRAZING ALLOYS

These 6% and 15% silver alloys have a melting point of 1185°F. and flows freely at 1300°F. These alloys should be used only on non-ferrous metals, particularly copper, brass and bronze. Here, the long melt range makes it important that the joint be heated to the flow temperature of the alloy before the alloy is applied.

In all cases where a long melt range alloy is used, care should be observed in heating to be certain that the work is brought up to the flow temperature of the alloy before application. Slow and extended heating with this type of alloy may result in liquidation, wherein the low melting constitutents of the alloy tend to flow out leaving behind a higher melting residue.

It is also recommended that a flux be used when brazing copper to brass or bronze.

SILVER BEARING SOLDER

This comparatively new solder in the field has proven to be a very successful item. Its high tensile strength and low melting point make it ideal for use in refrigeration applications.

Strong joints comparable to those of "45" silver braze are easily made.

Silver bearing solder flows at approximately 430° F. which puts it in the soft solder range *Figure 50B* low melting point effectively eliminates the chances of oxidizing the metal from too much heat. Silver bearing solders use an acid flux especially compounded for this type solder.

HOW TO USE

There are six simple but important steps which, if followed, will insure sound, strong, ductile, and leak tight joints:

1. Good fit and proper clearance.
2. Clean Metal.
3. Proper Fluxing.
4. Assembling and Supporting.
5. Heating and flowing the alloy.
6. Final cleaning.

GOOD FIT AND PROPER CLEARANCE

In silver brazing we are interested in flowing the alloy between closely fitted members, unlike other methods of metal joining such as high temperature brazing and welding wherein the filler metal is applied in quantity and generally in the form of fillets. By closely fitted members we mean parts having a joint clearance of approximately .0015 to 005. However, this does not mean that parts must be machined to exact tolerances. In the case of tubular members, an easy slip fit will be quite adequate and we recommend using a tubing cutter when cutting the tubes.

There are, of course, certain factors affecting tolerance which must be taken into consideration particularly as applied to tubular members. For example, in brazing a copper bushing into a steel sleeve we must take into account the fact that the copper has a greater coefficient of expansion than the steel and as it is to be the inner member of the assembly, we must allow a greater tolerance than if both pieces were steel. By the same premise, if the position of these parts were reversed, the copper becoming the outer member and the steel the inner member, we would probably want to allow a little lest tolerance than if both parts were of the same material. In general, tolerances should be considered in the light of the parts at brazing temperature rather than at room temperature.

Two reasons for maintaining a good fit on parts to be silver brazed are as follows:

Ease of Application:

In silver brazing we depend to a great degree upon capillary force to pull the alloy throughout the entire joint area. Excessively wide tolerances tend to break the capillary force and as a result the alloy will either fail to flow through out the joint or may flush out of the joint.

Corrosion Resistance:

There is also a direct relationship between the corrosion resistance of a joint and the clearance between members. Experience has taught that the greater the area of the alloy exposed to attack, the more rapid the corrosion; whereas the lesser the area, the slower the rate of corrosion, plus the tendency of the corroded area on the smaller joint to inhibit further corrective action.

CLEAN METAL

In silver brazing, as with all types of metal joining, it is most important to have joint surfaces clean. By this we mean they should be free from oil grease, oxide, scale and dirt. Attempting to braze contaminated or improper cleaned surfaces will generally result in an unsatisfactory joint. There are several reasons for this:

1. Silver brazing alloys will not flow over or bond to oxides.
2. Oily or greasy surfaces tends to repel the flux leaving bare spots which will oxidize when heated, resulting in voids and inclusions. Also oil and grease carbonize when heated, forming a film over which the alloy will not flow.
3. Dirt and scale prevent the proper application of the flux and also cause voids and inclusions.

There are many methods of cleaning, both chemical and abrasive. The type of contamination and the number of parts that are to be cleaned determine best the method to be used.

We recommend using emery cloth, for abrasive cleaning. Sand paper is not particularly recommended as it is not very durable and offers the possibility that the parts might become impregnated with silicon particles.

Care should be taken to avoid handling of the

joint areas after cleaning, as such a practice generally results in redepositing body oils and dirt on the surface.

PROPER FLUXING

Flux has three principal functions to perform:

1. It prevents the oxidation of metal surfaces during the heating operation, by excluding oxygen.
2. It absorbs and dissolves residual oxides that are on the surface and those oxides which may form during the heating operation.
3. It assists in the flow of the alloy by presenting a clean, nascent surface for the melted alloy to flow over.

In addition to the above, it is also an excellent temperature indicator since during heating Handy Flux passes through four-definite changes that can be readily observed. They are:

1. At 212°F. the water boils off.
2. At 600°F. the flux becomes white and slightly puffy and starts to "work".
3. At 800°F. it lays against the surface and has a milky appearance.
4. At 1100°F. it is completely clean and active and has the appearance of water. At this point a bright metal surface will be apparent underneath.

Handy Flux is a paste flux composed of potassium fluoride and other salts mixed with water and may be applied in several ways, such as: brushing, dipping, squirting on under pressure, etc. Which method of application is used, depends upon the type and number of parts to be fluxed, and ingenuity in developing a method of application best suited to the job.

Regardless of what method of application is used the prime consideration is to be sure that all joint surfaces of the parts to be brazed are completely covered with a coating of flux. If you desire to keep the areas surrounding a joint free from discoloration and oxidation, the flux should be applied to these as well.

To do a good job, the flux must wet the metal surfaces. If upon application it balls up, this indicates that the metal surfaces are not clean;

whereas when the flux easily wets the surfaces, it indicates that surface tension has been broken down and a good job of cleaning has been done.

Consistency of the flux and the proper amount to use are factors to be determined by the type of metal to be brazed and the length of time required for heating. If you will think of the flux in terms of a blotter which, instead of absorbing liquid. absorbs oxide, you will have little or no difficulty in making this determination. For example, if you were to join two light pieces of copper, copper oxide being comparatively soft and easy to remove, the heating cycle would be quite short and, as a result, the flux could be applied quite sparingly and in a fairly thin consistency. However, when· joining the same size pieces of stainless steel which has a very hard, refractory oxide, difficult to remove, you should apply the flux more liberally and a heavier consistency is recommended. Another example would be the brazing of two heavy pieces of copper. Though the oxide, as in the first case, is not difficult to remove, the heating cycle has been extended and additional flux of heavier consistency should be used. Always remember that flux is your insurance against oxide and that a specified amount of flux will absorb just so much oxide. Flux, like a blotter, once it becomes saturated, loses all its effectiveness.

It has been found that if practical, it is better to warm the flux (120°F. to 140°F.) prior to application as this helps to break down surface tension more readily than when the flux is at room temperature. In addition to assisting in the wetting action, the user will find that he will get a 25 to 40 per cent greater spreading as warm flux can and should be used in a thinner consistency than cold flux.

Though standard fluxing procedure calls for the coating of both surfaces of the joint, there are certain special cases which will not permit this, due to the danger of flux entrapment inside the system. In such cases, standard procedure is to assembly the joint and apply a fairly liberal amount of flux to the face of the joint. Upon heating a sufficient amount of flux will be drawn into the joint by capillary force to adequately clean and protect the surfaces. Such a procedure, however, can only be used where small joints are involved. For example, in the case of tubing this fluxing procedure is not recommended on joints above 5/8 in diameter.

On room air conditioners and domestic refrigerators we do not have any tubing greater than 5/8" and we should always be careful and not contaminate the refrigerant system. Be sure and insert the tube being joined first, then flux the outside of the joint.

ASSEMBLING AND SUPPORTING

Most refrigerant joints are so designed as to be self-supporting but at times a joint, such as a long tube, being brazed into a valve or fitting is not. If a long length of tubing is left unsupported, undue strain would be put on the joint and cracking or failure most likely would result in the valve or fitting.

Heating and cooling cause expansion and contraction, therefore, supporting of members is necessary. In high production work, permanent jigs and fixtures should be designed to fit the job. However, for small production work, almost any effective method of supporting can be used. In any case, it is always important to make sure a job is properly supported and that you have the smallest area of contract possible between the jig and the assembly to be brazed. Cumbersome and heavy jigs result in stealing the heat from the joint proper making it difficult, if not impossible, to complete the braze.

HEATING AND FLOWING THE ALLOY
(Refer to Figure 50—C)

There are many different methods of heating which can be used in silver brazing. Regardless of which method is used, it is important that both members of the joint be heated uniformly and both reach the brazing temperature at the same time and in as short a time as possible.

If a prest-o-lite tank is used we recommend at least a number 4 torch tip.

If oxy-acetylene is used we recommend a slightly reducing flame as illustrated in *Figure A*. A slightly reducing flame has a slight green feather and is rich in acetylene.

Some metals heat rapidly and some heat slowly **47**

Heavy sections of metal take longer to heat than light or thin sections. Also, some metals, such as silver and copper, having high heat conductivity carry off the heat into colder portions more rapidly than do other metals, such as 18–8 stainless steel, having low heat cond uctive. Therefore, when unequal thicknesses or masses are being joined, care must be used not to overheat the the thin section or underheat the heavy one. Also when joining a good conductor to a poor conductor, more heat will have to be applied to the good conductor.

NOTE: In all heating operations the flux should be used as a guide to the temperature of the parts being joined. If this practice is observed, even and correct heating will result regardless of the thermal conductively of the parts.

When both the members of the joint have reached the brazing temperatures, it is time to apply the brazing alloy to the heated surface. Capillary force then will pull it throughout the joint. It is desirable to maintain the heat for a short time to be sure that all flux and gases have been flushed out of the joint. This will also serve to obtain a good bond by insuring that both joint members are up to brazing temperature.

GENERAL HEATING INSTRUCTIONS MAKING HORIZONTAL JOINTS (Refer to Figure 50–C)

A. Use a low velocity bulbous oxyacetylene tip. Multiflame tips also work well.

Adjust torch for a slightly reducing flame, Figure (a).

C. Start heating pipe about 1/2'' to 1'' away from end of fitting, Figure (b). Heat evenly to get uniform expansion of pipe and to carry the heat uniformly to the end inside the fitting.

When flux on pipe adjacent to joint has melted to a clear liquid, transfer heat to fitting, Figure (c).

E. Sweep flame steadily back and forth from fitting to pipe, keeping it pointed toward pipe, Figure (c) the object is to bring fitting and

pipe up to an equal heat together for application of the silver brazing alloy.

F. When flux is a clear, fluid liquid on both fitting and pipe, pull flame back a little and apply firmly against pipe at junction between pipe and fitting Figure (c). With proper heating, alloy will flow freely into the joint.

MAKING VERTICAL DOWN JOINTS

A. In joining fittings to 3/4'' pipe or smaller, the entire joint can be brazed in one simultaneous heating operation.

B. After preliminary heating bring pipe and fitting to brazing temperature by wiping flame from back of bead of fitting toward pipe, Figure (e). When up to temperature, as indicated by clear, very fluid state of flux, apply silver brazing alloy and sweat it in.

MAKING VERTICAL UP JOINTS

A. Start with preliminary heating of pipe as before. When flux is completely clear and liquid, transfer heat to fitting and sweep back and forth from fitting to pipe, Figure (d). Be careful not to overheat pipe below fitting as this will cause alloy to run down pipe out of the joint.

B. When brazing temperature is reached, as indicated by flux, touch alloy to joint with heat aimed on wall of fitting to pull alloy up into the entire joint area.

CLEANING AFTER BRAZING

A. Immediately after brazing alloy has set, apply a wet brush or swab to joint, Figure (f) to crack and wash off flux. Use a wire brush if necessary.

TO TAKE JOINT APART

A. When necessary, joint can be taken apart as follows: Reflux entire joint area. Then heat entire joint uniformly to slightly above melting point of brazing alloy. Pipe can then be easily removed from fitting. Pipe and fittings that have been taken apart can be reused by following the preparation and brazing procedure given for original brazing. Always apply additional silver brazing alloy when rebrazing.

SILVER BRAZING

Experience over many years has proven that "silver brazing", when making repairs or replacing components, will give the most reliable leak proof joints. The use of silver alloy in brazing is very important, since the use of steel and copper tubing in the system makes it very difficult to obtain a good leak-tight joint with other soldering materials. With the use of 3145 Silver Braze, professional results can be obtained. Recently a new and economical silver brazing alloy known as Braz-Sil has been developed and requires no flux when used on copper to copper brazing. However where brazing is necessary between copper and brass, flux must be used. This brazing alloy has the same handling and characteristics as the more expensive Silver-Phosphorous brazing alloys.

To get the best leak-tight joints, the following must be adhered to.

1. Tubing must be snug fit with a clearance of no more than .006 inch, and one tube must overlap the other by approximately ½ inch. The capillary tube should be inserted into the dryer at least 1¼ inch, so solder does not run into the tube and seal the opening.
2. CLEAN THE TUBES to be joined. All surface oxides, paint corrosion, etc. must be removed to allow the alloy to flow properly. When using abrasive cloth for cleaning, be careful that dirt particles do not enter the tubing.
3. Before attempting to silver braze, the component or tubing should be supported so it does not pull away during heating or cooling. All tubings should be preformed and supported.
4. In applying the flux, care should be taken that the flux does not enter the tubing. It is a good policy to clean and flux the silver alloy also. Use flux around a joint to be unsoldered. Failure to use flux will make it difficult to unsolder and could result in carbonizing the tubings.
5. After solder is set, wipe the brazed joints with a wet cloth and remove any residue flux. If flux is left on, it could cover a leak that would appear only after vibration would loosen it from the tubing; the flux could also corrode the joint.

Recently a new and economical silver brazing alloy has been developed that requires no flux when used on copper to copper brazing. Where brazing is necessary between copper and brass however, flux must be used. This brazing alloy has the same handling and flow characteristics as the more expensive Silver-Phosphorus Brazing Alloys. It is available through your appliance parts jobber. Ask for Gemline Part Number 3108-8 Bras-Sil. The part number for the flux is Gemline 3107.

BRAZING EQUIPMENT

Since the term "hermetically sealed" describes a system that is welded together, tools must be available to open and reseal the refrigerant circuit. This requires the hot flame produced by **oxygen** and **acteylene** brazing and welding equipment.

This equipment is required for several reasons. The brazing material is a silver alloy and requires the heat of oxygen and acetylene, gauges and regulators, (one for oxygen and one for acetylene) hose and a welding torch with interchangeable tips. The equipment can be purchased from distributors of Refrigeration Parts and Supplies.

The other supplies needed with this brazing equipment includes brazing rod (for copper to copper joints) and flux.

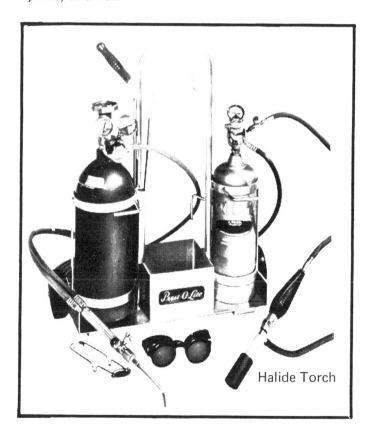

Halide Torch

PINCH-OFF TOOL

The proper repair of many hermetic systems requires the addition to the system of evacuation and charging stud. These are short pieces of ⅜" O.D. copper tubing (about 6" long) which are usually connected to the pinch off connections used to evacuate and charge the system during manufacture. After the repair is completed these tubes are pinched off again and **brazed shut** to reseal the system.

Quite often the pinch off is the weakest point in the system so it is of great importance that this pinch off be made with the proper tool *Figure 48 B* ˙. This tool must form a channel in the tubing thus making this section stiff. A tool that makes a flat pinch would form a weak section that has a tendency to break with very little vibration, *Figure 51* .

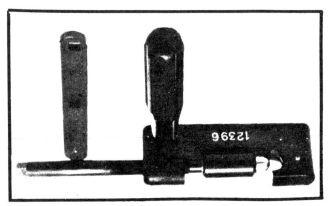

Figure 51

VACUUM INDICATOR

A good vacuum pump is of little use if we don't have and use a good vacuum gauge, capable of reading a high vacuum. Anyone that has tried to read the vacuum on a normal compound (Bourbon type) service gauge knows it is hard to tell whether the gauge is reading 28 or 29 inches vacuum. The problem is that these gauges are not accurate and the scale is too small.

A gauge rugged enough for field use, and yet accurate and economical is an absolute pressure gauge that will accurately determine when a safe vacuum has been reached (½ millimeter or below) and which is not affected by altitude or barometric pressure.

If it is desired to purchase a more accurate gauge use a Thermocouple Vacuum Gauge or a Gemline Vacuum Analyzer, *Figure 52* .

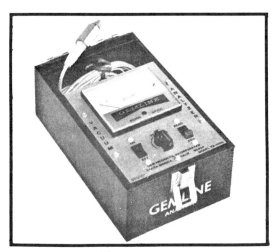

Figure 52

VACUUM PUMP

Since we have opened the system to make a repair, we have allowed air and moisture to enter the system. The only way to get this air and moisture out of the system is with a good two stage vacuum pump.

The pump required is **not** a rebuilt compressor but a pump designed as a high vacuum pump, *Figure 61.* The pump must be capable of pumping down to at least 5 microns and have a pumping speed of at least 20 liters per minute. These are actually small pumps and fairly easy to handle.

Many pumps are available and when purchasing these pumps it is a good idea to obtain the best you can afford. However this requires a good two stage vacuum pump if we want to remove much moisture. For instance at a 28" vacuum or (50,800 microns, absolute pressure) water boils at 101°F. Even under 29.5" vacuum or (12,700 microns, absolute pressure) water will boil at 59 degrees.

Knowing this we strive to reach 29.90-vacuum. (500 microns) absolute pressure for field repairs to hermetic systems.

Refrigeration engineers agree that to obtain a dry system the pump should be capable of pulling down within 50 to 100 microns.

WHY SUCH A GOOD VACUUM PUMP

Why is it necessary to use a pump that will obtain such a high vacuum? If we consider that water boils at 212° F. atmospheric pressure, and as we reduce the pressure on water it will boil at a lower temperature, it is easy to understand that we can evaporate water and remove it from a system as vapor by using a vacuum pump.

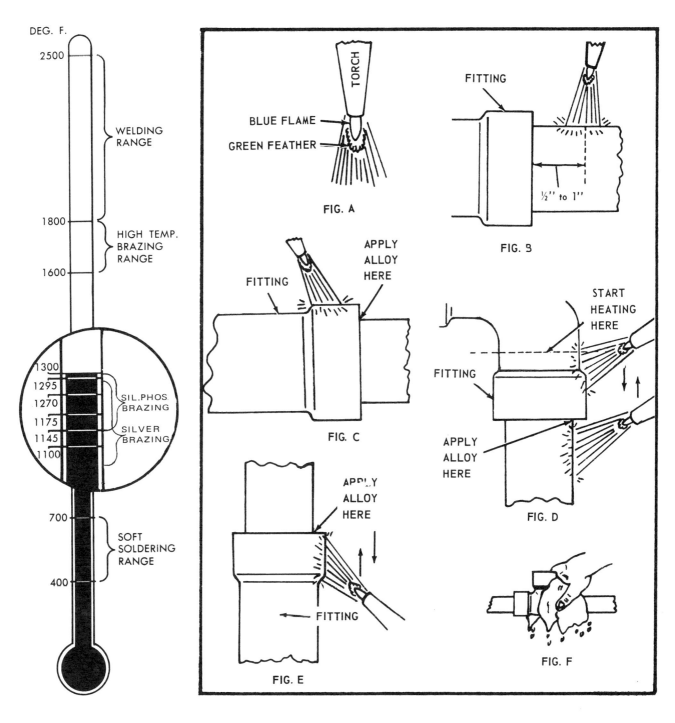

Figure 50B

Figure 50C

SWEDGING

When two lengths of copper tubing of the same size are to be joined without the use of a sweat or solder fitting, one of the two joining ends must be swedge or expanded enough so that the other end will fit snugly into the enlarged portion. See

Figure 48 . The joint is then completed by soldering or brazing in the same manner as has already been outlined under "Soldering."

The swedging process is accomplished by using

51

a specially designed punch, called a swedging tool, and a flaring block in which the tubing is clamped. The end of the swedging tool is called the pilot which fits loosely in the tubing. A tapered lead connects this pilot with an enlarged portion which is slightly larger than the outside diameter of the tubing. *See Figure 53*

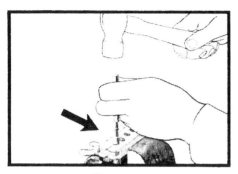

Figure 53

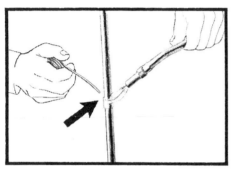

Figure 54

The tubing is now clamped in the proper size hole in the flaring block, and the flaring block, in turn, is secured in a vise. Place the tubing so it will extend above the flaring block a distance equal to the diameter of the tubing, plus 1/8 of an inch. The swedging tool is then driven into the tubing with a hammer until the tubing is expanded or swedged to a depth that is at least equal to the outside diameter of the tubing. A small amount of flux should be applied to the tapered lead of the swedging tool before it is inserted in the tubing. Do not use oil, as it will have to be removed prior to soldering, *Figure 54*

HANDLING REFRIGERANT

The Interstate Commerce Commission has set up definite rules governing the transportation of refrigerants with regard to the container (cylinder) in which the refrigerant may be shipped and the amount of refrigerant that each cylinder may contain by weight. These cylinders must be tested and approved for shipment of refrigerants because the pressure in the cylinder depends upon the temperature of the surrounding air.

The cylinder must be strong enough to withstand the pressure developed when the cylinder is exposed to the direct rays of the sun for long periods of time.

The amount of liquefied gas charged into a cylinder or drum must be determined by weight, and this weight must be checked by the use of proper scales after disconnecting the cylinder from the charging line.

Overcharging causes rapidly increasing pressure, which increases the temperature and may cause an explosion.

Normally-charged cylinders containing compressed gas, except those measuring less than twelve inches in length exclusive of the neck, must be equipped with one or more safety devices to prevent explosions when placed in a fire. A fusible plug is a solid metal plug filled with a special metal that melts at 165°F. Drums should never be heated above 129°F. for safety reasons. Extreme care should be employed in transferring the refrigerant from large cylinders to small service drums.

Make sure that the large valve containing the fusible plug is not overheated to a point where the metal will melt and permit complete discharge of the cylinder.

When charging service drums from large shipping cylinders, it becomes necessary to either have the service drum at lower temperature than room temperature, or to heat the large cylinder in order to establish the pressure difference necessary to cause the liquid refrigerant to flow into the service drum.

Shipping cylinders are always inverted when they are to be used for transferring refrigerant to smaller drums. This makes it possible to drain all the liquid refrigerant from the cylinder. When a service drum is to be filled, it must first be connected to the flexible charging line. Then it is placed on the scales and the weight of the empty drum noted. The valves can then be opened, and the additional weight as indicated by the scales will give the weight of the liquid refrigerant.

If many service cylinders are to be charged during the period of a day, it may not be adequate, and certainly not entirely safe, to heat the large drums. Instead, several methods of cooling the refrigerant prior to its entering the service drum have been developed. One such method of cooling is shown in *Figure 55* Several other commercial models are also available.

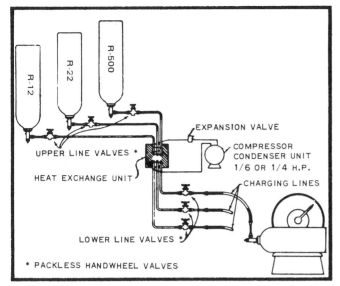

Figure 55

In this particular arrangement, the valves of the inverted supply cylinders are connected with copper tubing to the upper line valves which are placed on the wall or control board as near as possible to the cylinder outlet. These line valves connect to the heat exchanger unit, which in turn is connected to a set of lower line valves at the outlet of the heat exchanger unit. Flexible charging hoses are connected to the lower line valves, thereby making it easy to hook on service drums. The principle involved in this particular type of transfer arrangement is as follows; as the warm liquid refrigerant from the supply cylinder enters the heat exchanger, it is immediately cooled, after which it passes into the service drum. Once in the drum, the cold liquid will keep the pressure below that of the large cylinder, and the flow will continue. In passing through the heat exchanger unit, the warm

liquid from the large cylinder is continuously cooled by the operation of a small condensing unit. It requires only a very short time to fill a service drum with this method. Therefore, an attendant must be watching the filling operation and close the valve as soon as the scale indicates the correct amount.

It is well to mention here that in case the cylinder to be filled is not empty at the start, true empty weight must be used to prevent overfilling. All cylinders, except the smallest service drums, are fitted with safety plugs of fusible metal. These plugs will protect the cylinder when overheated, thus eliminating the possibility of a properly filled cylinder bursting due to excessive pressure. However, it is important that we all recognize the fact that a cylinder filled to more than its rated capacity will develop dangerous pressures before the melting temperature of the fusible metal is reached. It follows that safety plugs are ineffective as protective devices if the cylinder has been overfilled.

vice drum. This leaves only the vapor in the charging line as a loss in the transfer operation. The service drum valve can now be closed and the service drum disconnected.

One of the most common methods of transferring refrigerant from one drum to another is to pump a vacuum on the drum to be filled. Close the valve on the drum, and then disconnect it from the vacuum pump. Invert the supply cylinder so that the valve is at the bottom. Next, use a flexible charging line (as short as is practical) and connect one end of it to the service drum which is to be charged. Connect the other end of the charging line to the supply cylinder. Place the empty service drum on the scales and weigh it. Then, crack the valve on the supply cylinder and close it at once. The charging line will then contain a mixture of refrigerant and air. Loosen the connection very slightly at one end of the charging line and permit the refrigerant to escape slowly, so that the refrigerant will push the air out of the line at the same time. After the gas stops escaping, retighten the connection on the charging line and open the valve on the supply cylinder slightly. Then, open the valve on the service drum and you will hear the liquid refrigerant entering the service drum. Watch the scales, and when there is enough refrigerant in the service drum, close the valve on the supply cylinder. Warm the charging line in order to force the liquid refrigerant out of the charging line into the drum. Close the valve on the service

drum. Next, carefully loosen the connection on the charging line to let the gas in the line escape. Remove the charging line completely. The charging process is now finished, and the drum is charged with the desired amount of refrigerant.

Another method is to connect one end of the charging line to the service drum which is to be charged and then chill the drum by placing it in a pail of cold water Connect the other end of the charging line to the supply drum and purge the air as previously described. Close the valve on the drums and remove the charging line.

CHARGING CYLINDER

Many of the present day hermetic systems are equipped with capillary tubes for control of the refrigerant. This is a good feature that prolongs the life of the complete system, but these systems require closer tolerance of the refrigerant charge.

The use of a volume charging cylinder *Figure 56* is the most popular method of charging refrigerant into hermetic systems. It is important, however, to use a cylinder with a feature to compensate for variation in volume because of differences in pressure and temperature of the refrigerant.

Scales and precharged refrigerant cylinders can also be used if desired, however, the scales must be calibrated in increments of 1/4 oz.

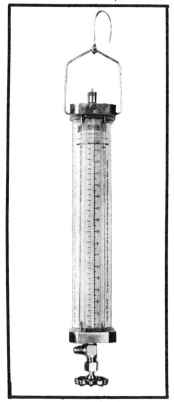

Figure 56

USING THE PIERCING VALVES

With the advent of the Hermetic sealed system, hermetic compressor manufacturers incorporated various specialized designs of, valves, fittings, ports, and process tubes for entry to the system. In order to service the system it required an extensive quantity of special tools to fit the many different compressor process fittings.

One such specialized tool, available in kit form was called a "Burglar Kit"! This expensive kit, composed of many pieces, was designed to adapt to different compressor fittings.

A simple, economical, standardized method of entry to the system was needed, and the line piercing valve was designed to meet the demand. *(Figure 58)* The line piercing valve requires no brazing or soldering installed. It is designed to clamp on the line and incorporates a permanent seal and a piercing needle for entry to the system.

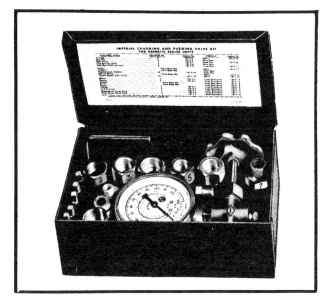

Figure 57

The piercing valve manufacturers instructions should be followed when installing a line piercing valve. Valves like the Gemline GPV-31 come complete with valve, inserts, wrench and instructions for easy installation.

1. If a piercing valve is used on the high side to get a gauge reading, it should be leak tested before the final charge of refrigerant is made. The piercing valve can be left on the system as a permanant part of the system, if no leaks are apparent.

2. Before a piercing valve is installed, clean and polish the tubing where it is to be placed. Look over the area, do not install the piercing valve on a bend in the tubing, or where the tubing is dented or kinked.

3. Install the piercing valve where it is convenient to work, and with the valve stem up.

Figures 59 thru *60* show the various types of Schrader valves and piercing valves. Also shown in *Figure 57* is the Hermetic Kit.

SCHRADER TYPE LINE ACCESS VALVES

The Schrader type access valve is an alternate design for entry to the sealed system. There are many variations of this type available.
Installation of this valve requires brazing or soldering of the valve to the line. A separate Schrader type valve core is inserted after the brazing process into the valve body, and is opened and closed by a depressor in the charging hose or depressor tool when servicing the system.

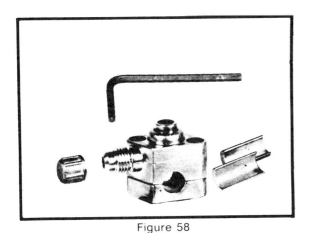

Figure 58

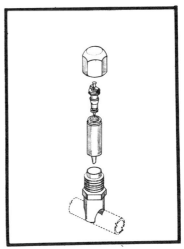

Figure 59

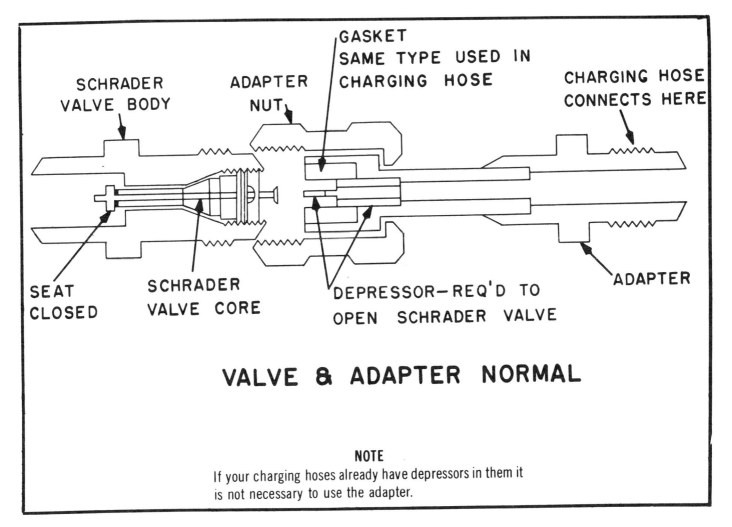

**SCHRADER
VALVE BODY**

**ADAPTER
NUT**

**GASKET
SAME TYPE USED IN
CHARGING HOSE**

**CHARGING HOSE
CONNECTS HERE**

**SEAT
CLOSED**

**SCHRADER
VALVE CORE**

**DEPRESSOR—REQ'D TO
OPEN SCHRADER VALVE**

ADAPTER

VALVE & ADAPTER NORMAL

NOTE

If your charging hoses already have depressors in them it
is not necessary to use the adapter.

Figure 60

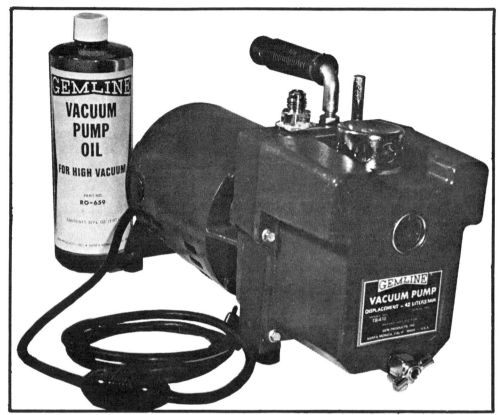

Figure 61   *Courtesy of Gem Products, Inc.*

ROTARY COMPRESSERS — CONDENSERS — EVAPORATORS

CLASSIFICATIONS

Rotary compressors vary somewhat in their interior construction, and are classified in regard to the method used for trapping and forcing the refrigerant vapors from the low-side of the system into the high-side. These two classifications are known as: (1) The rotary vane class and (2) the rotary squeeze class.

The rotating member in the rotary vane class compressor is provided with slots. Metal vanes or blades fit into these slots in such a manner that they are able to move in and out with considerable freedom without permitting the vapor to leak back around the vanes in the slots.

The rotating member in the rotary squeeze class compressor does not have slots, but depends instead on a single blade riding against the rotating

member to prevent the vapor from seeping back from the high-side into the low-side during operation.

ROTARY COMPRESSOR PARTS

Before we begin the discussion of the various makes of rotary compressors, it will be necessary for you to learn the names of all the parts and just how each individual part fits into the complete assembly of the compressor. These parts are entirely different from the ones used in the construction of a reciprocating compressor and should be throughly studied so that you will get a good mental picture of the entire compressor.

The parts common to all rotary compressors are: (1) The rotating member called the "rotor" or "impeller," depending on the type of compressor. (2) The shaft on which the rotor or impeller is mounted. In some compressors the rotor and shaft are made of one solid piece of steel rather than being assembled from two pieces. (3) The cylinder in which thr rotor or impeller moves or revolves. (4) The blade which contacts the rotor and divides the high-pressure side of the system from the low-pressure side, or the vanes which move up and down in slots in the impeller and make contact with the cylinder wall as the impeller revolves. (5) The housing or dome which encloses the entire compressor and provides space for oil and discharged from the compressor.

OPERATION OF ROTARY VANE COMPRESSORS

The rotary vane-type compressor is designed to use from 4 to 10 vanes in the impeller, depending on the size of the compressor and who manufacturers it. Smaller household compressors are made with 4 vanes in the impeller, but the larger capacity compressors have more vanes due to the larger diameter of the cylinder. The end plates on the cylinder carrying the bearings for the impeller have these bearings located off center, or eccentric to the center of the cylinder, so that part of the impeller is always in approximate contact with the cylinder wall.

Figure 62 shows the location of the impeller in relation to the inside of the cylinder. As the impeller revolves, the vanes are pressed against the cylinder wall by centrifugal force. This provides a space between each pair of blades. This space increases in sizes on the suction side of the compressor as the impeller revolves and permits additional refrigerant to enter the space.

As the next vane passes over the suction port, the vapor is trapped between the two vanes and is carried around to the discharge side of the cylinder where the space gradually decreases in size. This squeezes the vapor into a smaller volume which, in turn, causes the vapor to be discharged at a higher pressure than that at which it was taken in.

Several manufacturers of rotary vane compressors have used a number of different designs with respect to the angle of the blades. *Figure 63* shows a compressor in which the blades are set at an angle to the shaft rather than being set radial to the shaft as illustrated in *Figure 62*. In either case, however, the principle of compression is the same because the blades are riding against the cylinder wall and are being thrown out against it by centrifugal force. Therefore, the outer ends of the blades will follow the wall throughout each complete revolution.

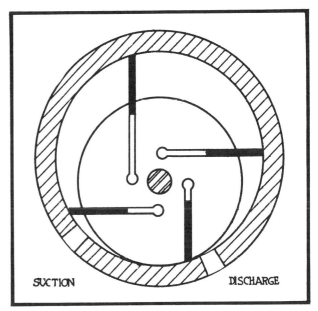

Figure 63

A rotary compressor is often driven by a split-phase motor through a direct drive. Split-phase motors have very little turning power (torque) when starting. Therefore, it becomes necessary to "unload" the compressor when the motor is started because the pressure difference between the high-side and the low-side would otherwise throw the full power requirements on the motor.

When the impeller is first started, the vanes are automatically forced back into their slots by the sudden motion of the impeller, thus allowing the

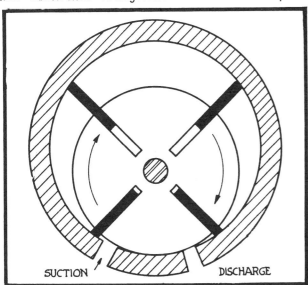

Figure 62

motor to bring the compressor up to speed before centrifugal force takes effect and throws the vanes out against the cylinder wall.

The compressor illustrated in *Figure 63* employs a unique method of unloading by having the blades set at an angle of approximately 30° from a center line through the impeller and shaft. The manufacturer claims that there is an advantage to this blade angle in that it will automatically unload the compressor in event liquid refrigerant or oil in any quantity enters the compression side.

In order to overcome the possibility of stalling the motor, the high pressure created by the incompressibility of the liquid, either oil or liquid refrigerant, will automatically force the vanes back into their slots and permit the liquid to be gradually worked through the compressor without stalling the motor.

Rotary vane compressors have also been designed with light coil springs placed behind the vanes to provide more positive contact with the cylinder walls during operation. As a rule, these springs are not necessary because there is sufficient centrifugal force exerted through the rotary motion of the impeller to hold the blades firmly against the cylinder wall at all times.

Most rotary compressors must be running at their designed speed before they will pump efficiently. Household compressors usually operate at motor speed (1725 R.P.M.) while commerical compressors may operate at from 1000 R.P.M. up to 1800 R.P.M. The vanes or blades do not make good contact against the cylinder walls until the compressor is revolving at about 1000 R.P.M. Therefore, the compressor throws no load on the motor until the compressor begins to approach its normal running speed.

As you learned from previous discussions of compressors, the vapor must be taken into the compressor at the evaporator, pressure and discharged from the compressor at condenser pressure. Refer to *Figure 64* and follow one complete revolution of the compressor to see just how it does compress the vapor. Suppose we start with the vane marked *1* and follow it around at progressive points in the direction indicated by the arrows on the impeller.

You will notice that this vane has just passed the discharge outlet and is approaching the suction side of the compressor.

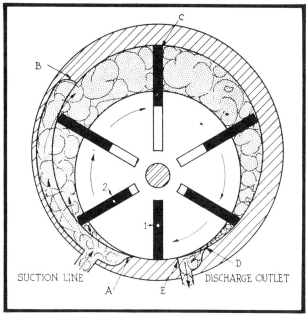

Figure 64

At point (A) it starts to pass over the slot in the cylinder wall where the vapor can pass into the compression chambers between the vanes.

From point (A) to point (C) the opening between vanes *1* and *2* is increasing in volume due to the offset position of the impeller with relation to the cylinder.

The moment vane *1* passes point (B) it traps the vapor between *1* and *2* and moves the vapor around until vane *1* comes to point (C) where compression starts between *1* and *2*.

The instant vane *2* passes point (D) it opens the enclosed space between the two vanes to the discharge outlet. All the vapor has been forced out of the confined space between vanes *1* and *2* when vane *1* returns to its original position (E). This same process is continuous between each two vanes whether the compressor is designed with 4, 6, 8, or 10 vanes. There is no pulsating effect of the vapor as it enters the suction inlet because the volume of the vapor between the impeller and cylinder is increasing gradually. Neither will there be any pulsating effect on the discharge side, provided, of course, the compression between any two vanes is equal to the condenser pressure when the leading vane approaches point (D).

In a rotary compressor we have a continous suction and discharge operation, whereas, in a reciprocating compressor, at the end of each compression stroke, the piston must return for a new charge of vapor. Therefore, a reciprocating type of compressor is only discharging vapor through one-half revolution of the crankshaft for each cylinder.

OPERATION OF ROTARY SQUEEZE COMPRESSORS

In the rotary squeeze class there are two types of compressors to be considered because there are two ways of mounting the blade which serves as charge sides of the compressor.

In the first type illustrated in *Figure 65* the blade is riding against the rotor (known as the roller in the Norge compressor). The blade is held tightly against the roller by a spring, as illustrated, and slides back and forth in a very finely machined slot that prevents the vapor from passing from the high compression side to the low suction side.

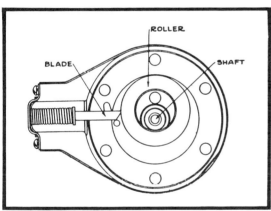

Figure 65

Notice that the center of the shaft is located in exact center of the cylinder. The shaft has an eccentric over which the roller slides. This makes the outside edge of the roller contact the cylinder through a continuous path as it rotates. Refer to *Figure 66*, which shows an eccentric shaft illustrated of the type that must be used in a rotary blade compressor. This eccentric shaft is very similar to those used in reciprocating compressors, but, as a rule, the eccentric does not have the throw required in a reciprocating compressor.

The second type of rotary squeeze compressor is illustrated in *Figure 67*. Here the blade is machined with a cylindrical end. This end fits into a socket on the rotor and is slide in lengthwise. Thus, the rotor and blade are attached on

a pivot which is realty is very similar to a hinge. As the shaft rotates the eccentric, it causes the contact point on the rotor to move around inside the cylinder while the rotor pushes and pulls the blade back and forth.

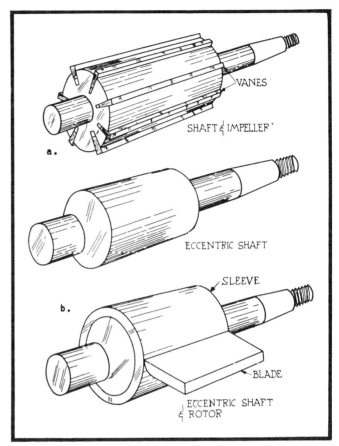

Figure 66

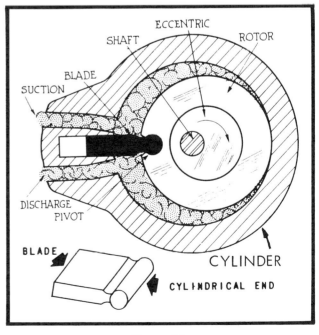

Figure 67

This type of compressor is referred to as a "semipendulum" compressor because the rotor does not revolve, but moves instead from side to side in relation to the blade and follows the blade as it moves back and forth in its machined slt.

The semipendulum type of compressor is usually driven at motor speed. Therefore, it need not be large in order to pump sufficient vapor to take care of the capacity required of a household refrigerator.

The parts, however, do have to be fitted to very close tolerances to prevent seepage of the vapor from the high-side to the low-side. This means the pivot point, the blade slot, and the point of contact between the rotor and the cylinder, as well as the covers over the ends of the cylinder, must fit perfectly with clearances usually less than one thousandth of an inch between the parts.

Trace the passage of the vapor through the semi-pendulum type compressor shown in *Figure 67* to be sure you understand how the vapor is compressed.

Low-pressure vapor enters the suction port and fills the entire space between the blade and the point of contact between the rotor and the cylinder wall on the suction side of the compressor. At the same time, the vapor on the discharge side is being squeezed out of the compressor because the rotor is constantly moving around in the cylinder in the direction indicated by the arrow on the eccentric.

The instant when all the vapor has been discharged, the suction port is momentarily closed by the rotor until the rotor moves past the opening and again starts to take in low-pressure vapor. A certain point on the rotor does not change its relationship to the point of contact on the cylinder wall, because the rotor is not free to revolve as it is in the Norge compressor illustrated in *Figure 65*.

Notice that the roller is nothing more than a ring mounted on the eccentric; hence, it does have a tendency to move around or change positions in the cylinder, but it does not revolve as rapidly as does the eccentric.

The rotary compressor is a continuous pump rather than an intermittent one as is the reciprocating compressor. You can see in the diagrams that vapor will be taken into the suction side at the same time other vapor is being discharged on the discharge side. Therefore, this compressor is capable of handling a large volume of vapor compared to its relatively small size.

Now, study the complete cycle of the Norge Rollator Compressor to see how it operates. Referring to *Figure 68*. Notice that the end of the housing is cut away to show the working parts of the compressor at different stages of one revolution.

The first diagram shows the roller covering the suction port, and the blade is pushed back into its slot flush with the cylinder wall. At this point, the cylinder is full of vapor and at the same pressure as on the suction side of the system.

The second diagram shows the roller in a position in which it has reduced the volume slightly on the discharge side and has started to take in fresh vapor from the suction line. The vapor is gradually being compressed into a pressure equal to that in the condenser and in the space at the top of the housing from where the vapor actually leaves the compressor.

The third diagram shows the roller in a position in which half the volume inside the cylinder is under half pressure and the other half is under low pressure. At this point during the compression stroke, the pressure on the discharge side has become sufficiently high to be about equal to the discharge pressure, and the discharge valve is just ready to open.

The discharge valve is located in the small slot in the cylinder housing directly above the blade. This valve can be seen better in *Figure 69* It consists of a small disc arranged in such a manner that the pressure built up in the cylinder will open it. After completion of the compression stroke, the higher pressure above the discharge valve will automatically push the disc back over the seat.

The fourth diagram in *Figure 68* shows the roller just about at the completion of the compression

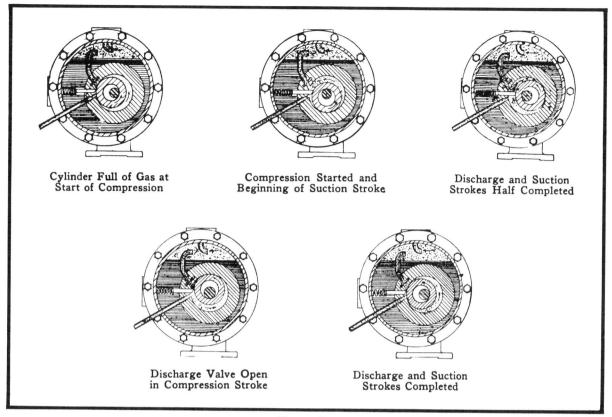

Cylinder Full of Gas at Start of Compression

Compression Started and Beginning of Suction Stroke

Discharge and Suction Strokes Half Completed

Discharge Valve Open in Compression Stroke

Discharge and Suction Strokes Completed

Figure 68

stroke, while on the other side of the blade, the suction stroke is also nearing completion.

In the fifth diagram, the roller has returned to its original position and the cylinder is again full of vapor and ready to be compressed.

SHAFT SEALS

All of the rotary compressors we have discussed up to this point have the same type of shaft seals as are used on reciprocating compressors. However, the seals on rotary compressors are located on the high-side of the system instead of on the low-side as is the case with reciprocating compressors.

The high-pressure vapor is discharged into the housing surrounding the compressor. Since the shaft must extend through this housing, it follows that the seal must be under high pressure at all times.

There is one advantage in having the seal on the high-side; it eliminates the possibility of air being taken into the system around the shaft in the event of a seal leak. This is particularly ture of systems in which the pressure on the

low side is below atmospheric pressure, as is the case with certain refrigerants, and with all the common refrigerants when extremely low temperatures are desired.

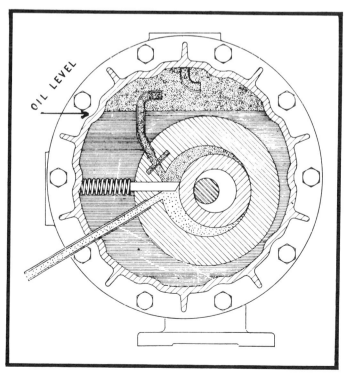

OIL LEVEL

Figure 69

Another advantage in having the seal on the high-side is the fact that the compressor housing also contains the oil for lubrication and, therefore, oil is surrounding the seal nose at all times and giving it proper lubrication.

If you will again refer to *Figure 69*, you will note that the oil level is entirely above the compressor cylinder. Since the seal is on the shaft, it is well submerged in oil at all times. Any oil that may be pumped through the compressor is discharged immediately through the discharge pipe above the oil level and falls back into the main body of oil.

It is very important to maintain the proper amount of oil in a rotary compressor because the oil serves as a seal between the rotor and the wall of the cylinder. If this seal did not exist, the capacity of the compressor would be decreased because some of the refrigerant gas would seep back to the low-side.

Good oil circulation must also be maintained for proper lubrication. Becuase the parts themselves

are fitted so closely that they would only last a few minutes if the oil did not form a film on their surface at all times.

In addition to having belt-driven rotary compressors, the same type of compressors are also used very extensively in sealed units where the motor itself is built into the same housing as the compressor. This design, of course, eliminates the necessity of having a seal on the compressor shaft, because leakage around the bearings will only go from the high-side to the low-side of the system instead of to the atmosphere.

Figure 70 shows the design of a sealed system used in the Norge refrigerating unit. At the bottom of the motor-compressor shaft is a disc, or impeller, that forces the oil to travel through a passage to the upper oil reservoir. From here the oil flows down into the moving parts of the compressor, and is finally carried up through the grooves indicated by the slanting arrow across the shaft and falls into the upper oil reservoir.

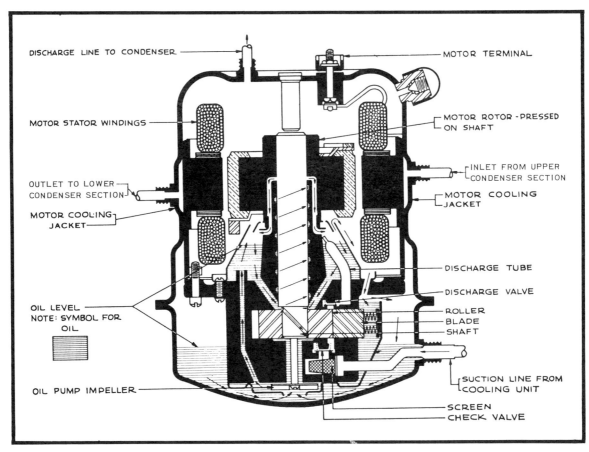

Figure 70

The construction of this compressor is essentially the same as that shown in *Figure 69*, but instead of the suction and discharge ports coming through the cylinder wall, they are located in the bottom and top cover plates of the cylinder. The suction port is, of course, on the suction side of the blade, while the discharge port is on the opposite side which can be move conveniently seen by referring to *Figure 70*.

The low-pressure vapor enters the compressor at the point marked "suction line from cooling unit" and passes through a screen, then up through a check valve before entering the compressor.

Up to this point we have not discussed check valves, but it is well to say here that all rotary compressors require check valves on the suction side to prevent the vapor on the high-side from leaking back through the compressor into the low-side during the off cycle.

The parts do not fit sufficiently tight to prevent leakage while the machine is at idle, but very little vapor seeps back while it is in operation. Therefore, you will always find a check valve on the suction side of a well-designed rotary compressor.

In the compressor illustrated ir *Figure 70* , the check valve is nothing but a disc mounted over the suction opening. The difference in pressure between the vapor in the suction line and the compressor pressure raises the valve and permits the vapor to flow into the suction side of the compressor.

The compression cycle in the Norge sealed unit compressor is the same as that of the belt-driven one in *Figure 68*. After the high-pressure vapor leaves the compressor, it passes through the discharge tube to the upper portion of the compressor-motor housing. The high-pressure vapor then leaves the housing through the discharge line, and enters the condenser where it is cooled.

In this particular design of a sealed system, there are no means by which the heat of the motor itself can be dissipated directly to the atmosphere. Therefore, it becomes necessary to provide some cooling in order to keep the motor from running too hot.

If you look at the diagram of the Norge sealed unit in *Figure 71*, you notice that the condenser is divided into two sections.

After most of the heat has been removed from the high-pressure liquid refrigerant in the upper section of the condenser, the last coil of this section is passed through the jacket surrounding the motor. (The inlet and outlet connections to this part of the condenser are clearly shown in *Figure 70*).

While the liquid refrigerant is circulating through the jacket, it is absorbing heat from the motor and reducing the motor temperature. The refrigerant is then returned to the lower section of the condenser where the added heat is being removed and the remaining vapor is being condensed into a liquid. More condenser area is required when a part of the condenser is used to cool the motor, but it proves very effective in maintaining a lower operating temperature of the motor and thereby increases its life.

This is also a very positive method of cooling because they are actually circulating liquid refrigerant around the motor, and are making use of the latent heat of vaporization by evaporating a certain amount of the liquid as it comes in contact with the hot seal poles. The whole interior of the motor assembly is exposed to the high-pressure vapor which, as you learned before, is also at a high temperature.

A number of other methods are also being used to cool the motor in hermetically sealed units. One method is to use the oil circulating system in which the oil itself is made to circulate through a few rows of extra coils in the condenser and back into the motor-compressor compartment.

Another method consists of using the cool suction vapor to cool the motor. In this case, the entire housing surrounding the motor and compressor serves as a passage for the low-pressure suction vapor coming directly from the evaporator. Thus, the cool vapor tends to pick up additional heat created in the motor by the electric current. This heat is then carried through the compressor and dissipated by the condenser, together with the heat already present in the suction vapor.

LUBRICATION

Inasmuch as the lubricating oil in rotary com-**63**

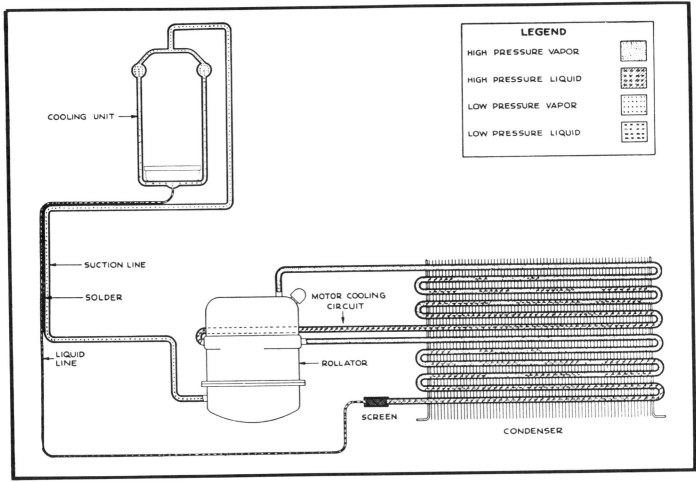

LEGEND

HIGH PRESSURE VAPOR

HIGH PRESSURE LIQUID

LOW PRESSURE VAPOR

LOW PRESSURE LIQUID

COOLING UNIT

SUCTION LINE

SOLDER

LIQUID LINE

MOTOR COOLING CIRCUIT

ROLLATOR

SCREEN

CONDENSER

Figure 71

pressors is exposed to high-pressure, high-temperature refrigerant vapors, manufacturers recommend oil of somewhat higher viscosity

In addition to the oil being thinned out by the high temperatures, it must also be remembered that when oil is exposed to high-pressure refrigerants, it will absorb more of the refrigerant. This has a tendency to thin oil.

Many manufacturers of refrigeration units also design the lubricating system in such a manner that the oil is cooled through either an auxiliary condenser or through passages surrounding some of the cooler parts of the unit. The purpose of this design is to prevent excessively high temperatures from being developed in the oil and, at the same time, delivering cooler oil to the moving parts.

SEALED UNITS

The sealed unit was developed in the late twenties for domestic use. Each year, more and more manufacturers changed over to this type so that today all manufacturers of domestic refrigerators are making sealed units.

In the sealed unit, the compressor shaft seal has been eliminated. This is accomplished by placing the motor and compressor on the same shaft, which makes the compressor directly driven. A housing is then placed around both the motor and the compressor and welded into one piece. This forms a motor-compressor unit which is hermetically sealed.

If the motor compressor housing is constructed with flanged joints, using bolts and gaskets for assembly, it is known as a semi-sealed unit or as a field serviceable unit.

A further classification is made of sealed units in regard to the manner in which the unit is connected to the evaporator and condenser. Under this classification we have three types: (1) Completely sealed units. These units have no service connections of any kind, either on the compressor or on the lines. (2) Hermetically sealed units.

These are units with opening into the one unit, either on the compressor housing or on the refrigerant lines. This opening is usually on the high side. The opening is such that a special tool must be used to open the service valve. (3) Semisealed units. These are complete units on which the motor-compressor may be removed without removing any other part of the system. On some semisealed units, regular service valves are installed and, on others, special service valves are used which require a special tool kit in order to make them serviceable by a serviceman.

The special tool kit mentioned above is illustrated in *Figure 72* and is known as a Service Valve Kit for sealed units. The kit consists of a valve body, complete with stem and packing, a gauge connection on the valve body, and several adapters and wrenches that make it possible to adapt the valve to practically any make of unit. With this kit it is possible to take pressure readings, purge or add refrigerant, or evacuate any of the hermetically sealed or semisealed units on which this type of special service valve can be used.

Figure 72

Other means of obtaining access valves as shown in *Figure 73* and *74*. Special adapter valves are available to connect gauges and charging lines to the factory charging stub by cutting off the end of the tubing stub and clamping the valve onto the remaining stub. After charging, the tubing stub is pinched, the valve removed, and the stub end brazed to seal the system.

Sealed units were first developed for domestic refrigeration, and it is in this field that they have

had their greatest application. Semisealed units are also used in commerical application and are being built in sizes up to 100 horsepower. The larger units are similar in construction to the smaller ones, except for the bore and number of cylinders in the compressor. A different type of motor is also used. We will first discuss the small domestic units.

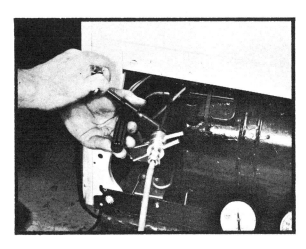

Figure 73

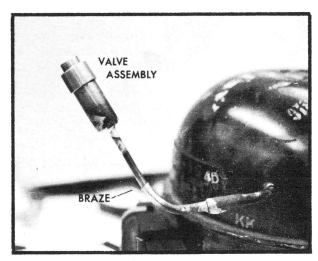

Figure 74

COMPRESSORS IN SEALED UNITS

Since the compressor is driven at motor speed, one of the first things to consider is vibration at full motor speed. One means of keeping the vibration at a minimum is to use a rotary compressor. *See Figure 70*

ROTARY COMPRESSORS

Two types of rotary compressors are being used: (1) the rotary vane-type and (2) the rotary-squeeze type. Information about the construction of these compressors was given in an earlier section.

When a rotary compressor is used in a sealed unit, the motor-compressor assembly is mounted vertically. That is, the motor is mounted above the compressor so that the drive shaft is in a vertical position. When the compressor is mounted in this manner, it is near the oil pump, and positive oiling is obtained in either of two ways: (1) by an oil impeller on the end of the shaft forcing the oil to the upper oil reservoir or (2) by utilizing the fact that the oil is under the discharge pressure of the compressor. In the second oiling method there is an oil control which opens directly into the cylinder at the point where compression begins. The oil is discharged along with the gas into the dome of the unit. From here the oil falls into a reservoir where it drains by gravity into the shaft spiral and is carried by this spiral to the top bearing.

Rotary compressors are used on both hermetically sealed and semisealed units. Because the oil in rotary compressors is exposed to the high-temperature, high-pressure refrigerant, it will thin out. Several methods of cooling the oil have been tried. One method is to place an extra row of tubes in the condenser and circulate the oil through this part of the condenser, thus obtaining cooler oil to introduce to the bearing surfaces. Another method of cooling has been employed by which the refrigerant is cooled in the upper part of the condenser and then returned to a jacket surrounding the motor.

The refrigerant picks up the heat produced by the electric current flowing in the motor windings and also some of the heat produced by friction in the moving parts. The refrigerant then returns to the lower parts of the condenser where removal of heat continues and condensation finally takes place.

RECIPROCATING COMPRESSORS

Reciprocating compressors are being used by many manufacturers of sealed units. *See Figure 75*. There are three major types of reciprocating compressors, namely, the true crank types, the eccentric type, and the Scots yoke type. We will not attempt to describe them in detail here since

they have been discussed elsewhere, but we will discuss each type enough to give you a clear picture of how the compressor is mounted in a sealed unit.

In the true crank-type compressor, the alignment of the drive and crank in relation to the connecting rod and piston pin must be held within very close limits in order to eliminate noise when the compressor is under load and running at motor speed. Rebuilding of the motor-compressor unit in the service ship is not advisable unless one has had special training in machine shop practices. Usually, the drive shaft on this design of unit is mounted vertically so that the piston travels horizontally. Lubrication is either force-feed or the splash type.

By keeping the stroke of the piston short, the reciprocating compressor can be operated at motor speed without producing noticeable vibration.

Eccentric Drive Compressors

Several manufacturers are using the eccentric drive compressor with good results. It is easy to disassemble and reassemble because many variations are possible regarding the alignment of the parts. This compressor is built with the drive shaft in either a horizontal or vertical position. If the design is of the vertical type, the compressor is located below the motor and is partially submerged in oil. All exposed parts are splash lubricated, while the main bearings have forced lubrication. The pressure is produced by centrifugal force acting on the oil as it passes through the hollow shaft.

The Scots Yoke Compressor

The Scots yoke compressor is so designed that flexibility is secured in the movement of its parts in any direction without bending, even though the motor is not in perfect alignment with the piston. *See Figure 76*. The motor and compressor are mounted either in a horizontal or vertical plane. In the horizontal type, the cylinder is either above or below the drive shaft. In the vertical type, the compressor is placed either above or below the motor. Each manufacturer has tried one or both of these locations for the compressor.

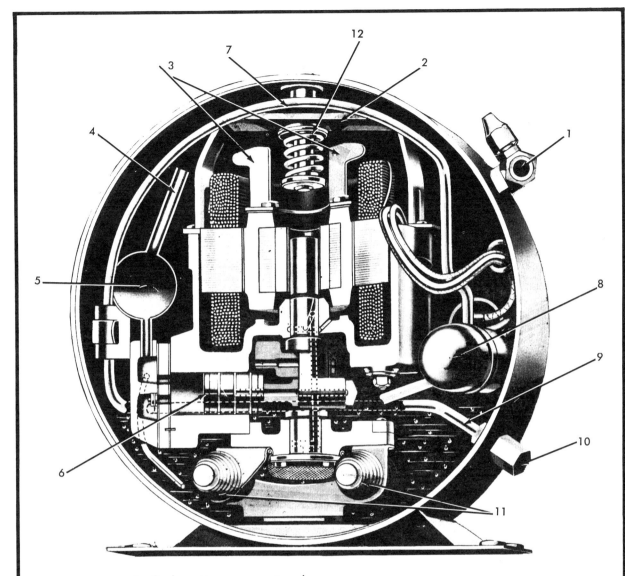

1. Cool suction vapor enters here.

2. Velocity of vapor is reduced here, permitting oil to drop by gravity into oil reservoir.

3. Impeller blades on motor shield cool vapor to inner shell surface, carrying internal heat to shell for cooler operation.

4. Small-diameter tube picks up only oil-free dry vapor.

5. Suction muffler absorbs sound of suction valve pulsations.

6. Vapor is compressed in cylinder by piston.

7. Compressed "hot" discharge vapor in tube passes through cool suction vapor for maximum pre-cooling effect.

8. Discharge muffler absorbs sound of discharge and compression.

9. Discharge vapor in tube passes through oil reservoir for additional heat transfer and increased operational efficiency.

10. High pressure refrigerant vapor leaves compressor via this discharge outlet.

11. Twin stabilizer springs.

12. Torque and suspension spring.

Figure 75

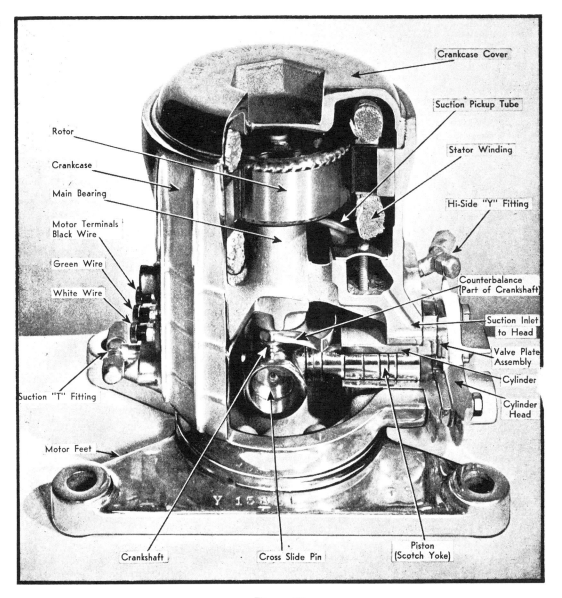

Figure 76

Most reciprocating piston compressors, regardless of the design, make use of the cool refrigerant vapor to cool the motor windings. This is accomplished by introducing the low-pressure vapor from the suction line directly into the housing where it surrounds the motor and compressor cylinder. Since this vapor is at a relatively low temperature, it absorbs the heat from the motor windings rapidly, thus keeping the operating temperature of the motor within safe limits. This also means that the housing of the motor-compressor is not subjected to high pressures during operation of the unit.

The suction intake to the compressor is located in the low-pressure area and at a point where oil is not likely to enter. In many units, the air gap or space between the field coils and the rotor of the motor is utilized as an oil separator. In these units, the suction intake is located above the rotor of the motor. In other units, the openings in the rotor are used as oil separators. These units have the suction intake just below the rotor and near the main bearing.

CONDENSERS

When sealed units were first introduced, natural-draft, bare-tube condensers were used. Any air-cooled condenser without an air circulating fan is natural draft condenser. This design has the advantage of being noiseless since there is no fan to move the air. A disadvantage is found in the fact that air flow to the condenser can be easily blocked. Many times after a unit was installed, the customer would partialy enclose the unit and thus increase the running time.

Another design of natural draft condenser is the popular finned tube type. In later models, a flue is placed behind the cabinet to direct the warm air away from the condenser. This prevents the owner from blocking the air flow by placing the cabinet too close to the wall. Usually, this type of condenser is mounted below and to the rear of the food compartment in the coolest air of the room.

The third design of natural draft condensers used with sealed units is the plate type. This condenser is mounted on the back of the cabinet and forms its own flue. It covers nearly the entire back of the cabinet. A plate-type condenser is made by taking sheets of metal and stamping them out with a definite channel so that when two of these sheets are placed together and the edges welded, a path for the refrigerant is established between the two sheets, and the condenser looks like a plate of corrugated sheet metal.

Natural draft condensers have a large mass of metal that will absorb heat while the unit is running and give up this heat after the unit turns off. As a result, the average head pressure is lower than would be expected in this type of condenser. The operating pressure may even be lower than that in a small, forced-draft condenser. This condition exists because of the short time the machine is running during each cycle, and, of course, anything that will increase the running time of the unit will cause the head pressure to increase while the absorbing effect of the condenser is reduced. As a result, the compressor and motor must be of rugged construction and able to withstand the added load. Another factor to be considered is that less cleaning of the condenser is required because dirt is not forced against the condenser by a fan. Noise as a source of trouble is also eliminated by not using a fan. Natural-draft condensers can only be used on the smaller units and where the load conditions do not vary a great amount.

Another condenser used on air-cooled, sealed units is the forced-convection condenser. This condenser has been used by several manufacturers, and the usual design has been of the tube-and-fin of suitable size and mounted in the motor compartment. This type of forced convection condenser has a separate motor and fan assembly mounted in such a manner that it will force air through the condenser and in this way cool the refrigerant. A few models with plate condensers have also used a motor-fan assembly.

REFRIGERANT CONTROLS

The refrigerant flow controls used in domestic, sealed units are the capillary tube and the expansion valve. The capillary tube is known by such names as capillary line, impedance tube, and restrictor. Both of these refrigerant flow controls are sensitive to changes in the charge of refrigerant. Great care should be taken when charging a unit to avoid getting an over charge or undercharge of refrigerant in the system. The result will be longer running time and lowered efficiency of the unit. In commerical, sealed units, any type of refrigerant flow control may be used.

EVAPORATORS

One of the popular shaped evaporated used in domestic seal units is of the U-design with refrigerated shelves. A study of *Figure 77* will show why this design is popular. The liquid refrigerant can be fed into the evaporator, either at the bottom or anywhere along one side. In this manner, the entering liquid will agitate the liquid already in the evaporator and increase the speed of vaporization.

The header or accumulator collects the vapor from the boiling refrigerant and passes it into the suction line. The accumulator also allows for the expansion of the liquid in the evaporator when it is permitted to warm up. An accumulator is necessary to prevent liquid refrigerant from entering the suction line when a warm unit is started. The shelves or shelf are refrigerated to increase the speed of freezing ice cubes. A point to remember about this is the fact that the top trays will be slower in freezing than the trays located lower in the evaporator. The ice cube trays placed on the bottom of the evaporator will freeze in the shortest time.

Another design of evaporator used by some manufacturers is the continuous tube type. The tubing may be wound around a shell which forms a freezing compartment and is enclosed by a door placed in the front. This design is coming into favor for storage of frozen foods. *Figure 78*

69

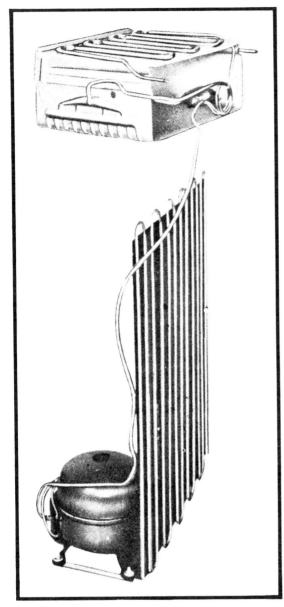

Figure 77

The tubing may also be placed on one side of heavy-gauge sheet metal which serves as a shelf. Continuous tube evaporation must also have provision for the vaporizing of the refrigerant. The tubing must therefore be increased in size to form a header or accumulator. This accumulator is formed at the outlet of the evaporator.

Aluminum roll-bonded evaporators are now used in a great number of domestic refrigerators. Two sheets of aluminum are pressed together under high pressure which bonds them together. One sheet is treated with a material which will not adhere to the other sheet with the shape and design of the refrigerant circuit. The bonded sheets are then bent into the final shape of the evaporator. Air pressure is connected to the inlet connection

and the refrigerant circuit is formed by the pressure applied through the unbonded area. The coil is held in a mold form during this operation to shape the expanding aluminum into its proper design, shape, and circuit for containing the refrigerant.

Figure 78

MOTORS IN SEALED UNITS

We have followed the refrigerant cycle from the compressor through the condenser, refrigerant flow control, evaporator, and the suction line. The next item we need is a driving force for the compressor. This force is obtained from an electric motor. Since the motor is located in the same housing as the compressor and is exposed to refrigerant and oil, a few special problems had to be overcome in order to produce a suitable motor for a sealed unit.

One factor which had to be considered was the materials used for insulation of the motor windings and in the stator slots. These materials had to be able to withstand the effects of oil and refrigerant. Also, if the winding insulation were of cotton, it had to be free of lint so that screens would not be plugged in any part of the system.

Another difficulty that had to be overcome in the sealed unit was arcing when the motor started. Since sparks would cause the oil to carbonize, or cause a breakdown of the refrigerant, motors with brushes could not be used. This meant that repulsion-start, induction-run motors, as well as d-c motors, could not be used in sealed units. The only motors that could be used in sealed units were capacitor and split-phase motors for domestic refrigeration units and squirrel-cage motors for larger units. These three motors are so designed that the starting equipment may be located outside the sealed unit housing so that any sparks produced when the circuit is opened will have no adverse effect.

One or two manufacturers make sealed units to be used in areas that are furnitured with only direct current. These units have an additional mechanism that converts direct current into alternating current so that the motors used to drive the compressors of these units can still be a-c motors. The device used for conversion is known as a rotary converter. Capacitor motors used with sealed units are of two types, namely, two-value capacitor motors and capacitor-start, induction-run motors. The two-value capacitor type is divided into two designs: (1) the capacitor transformer design, and (2) the dual capacitor design.

In the capacitor transformer design, both the main and the auxiliary windings carry current all the time the motor operates. The auxiliary winding is in series with the main winding of an auto-transformer. The capacitor is the load on the auxiliary or secondary winding of the auto-transformer. The amount of wire used for the main winding is varied by the use of a magnetic switch connects a tap on the transformer to the auxiliary winding. *Figure 79* shows a schematic diagram of this type of motor. You will note that by moving from one tap to another on the transformer, the amount of winding in the main is increased or decreased, depending on which tap is used. This means that a high voltage is impressed across the capacitor during the starting period, while the voltage impressed across the capacitor during the running period is much lower, usually two and one-half to three times the line voltage. The starting voltage may be as high as 600 volts. These high voltages are confined within the steel shell which encloses the capacitor and transformer.

The dual capacitor motor has two capacitors. One is known as the running capacitor and is in series with the auxiliary winding at all times. The other

is known as the starting capacitor and is used only when the motor is connected to the auxiliriy winding of the motor through a magnetic relay which opens the circuit when the current flow reaches a predetermined value. *Figure 80* shows a schematic diagram of the wiring of this type of motor. A study of the diagram will show that the two capacitors are in parallel with each other. The running capacitor is in use all the time the motor is running.

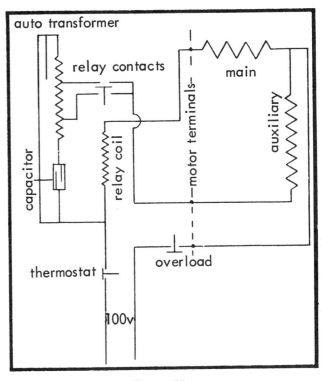

Figure 79

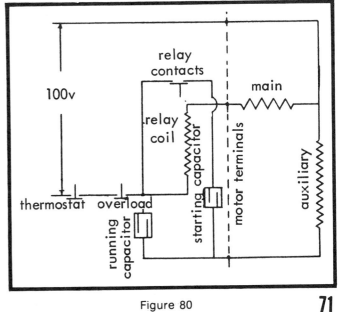

Figure 80

The capacitor-start, induction-run motor has only one capacitor. This capacitor is in the circuit for only a few seconds during the starting period of the motor. The only difference in the design between this motor and the open-type capacitor motor is the elimination of the centrifugal switch and placing the capacitor in a separate casing that is not connected to the motor housing.

Capacitor-type motors have the advantage of being able to start under load and are very quiet in operation. Electrolytic capacitors are usually used for starting purposes on these motors. Most manufacturers use an oil capacitor for the running capacitor.

The split-phase motor has probably been used more than any other in domestic, sealed units because the initial cost is lower, and it is also quiet in operation. The motor derives its name from the fact that the stator has two separate windings that are 90 electrical degrees apart. Single-phase current is fed through both of these windings during the starting period. In order to get the rotor to move, the current in the secondary winding must lag behind the current in the main winding. This is accomplished by adding resistance to the secondary circuit; hence the name "split-phase."

On older style, split-phase motors, the splitting was accomplished by using an external resistor added to the secondary circuit, but in later models the resistor is not used. Instead, the same result is achieved by a special design of the motor windings. A split-phase motor has low starting torque, and, as a result, some type of unloading device must be used on the compressor.

UNLOADING DEVICES

Sealed units that require a motor larger than one or two horsepower are usually built with a three-phase motor of the Squirrel-cage type. This type of motor is very simple in construction and requires no special starting devices. It uses an ordinary temperature or pressure control as a pilot control for a magnetic across-the-line starter.

One of the more common methods of unloading the compressor is to use a small diameter liquid line as a throttling device. As a result, the line serves two purposes: (1) It performs as the refrigerant

flow control while the unit is running. (2) It acts as a pressure equalizer while the unit is on the OFF cycle. We have already discussed the cappillary tube as a throttling device in a previous section. Because the capillary tube also serves as a pressure-reducing device, the high-pressure liquid in the condenser will continue to pass through the line into the low side during the OFF cycle, and, as a result, the back pressure and the head pressure are equal, or nearly so, when the compressor is started again. We know that little power is required to drive a compressor when there is no difference in back and head pressures. Hence, a split-phase motor is able to start the compressor.

Magnetic unloaders

The Westinghouse Company uses an unloader on its sealed unit that works with a magnetic coil. *Figure 81* is a cutaway of this unloader, showing its general construction. Here is another use for the solenoid valve, and in this case, gravity does not assist in closing the valve. Instead, a small spring is used to close the valve.

The unloader is located outside the cylinder wall and opens or closes a port that extends into the inner wall of the cylinder near the top, or into the clearance space of the cylinder.

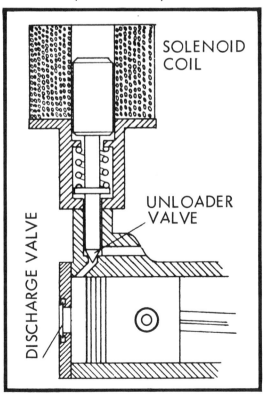

Figure 81

In order to give the solenoid coil enough power to operate the valve needle, it is wired into the common lead of the motor so that the current from the main winding and the current from the secondary winding pass through the solenoid coil. *Figure 82* is a schematic diagram of the wiring of this unit. The unloader works because the compressor is located in the low-pressure space. When the motor starts and the piston travels up on its compression stroke, the unloader valve is open and the refrigerant vapor is pushed back into the crankcase of the compressor which is under low pressure. As a result, the load on the motor is quite small. When the motor has gained in speed to a point where it can carry the full load of the compressor, the unloader valve closes and the compressor begins to operate the same as any other reciprocating compressor.

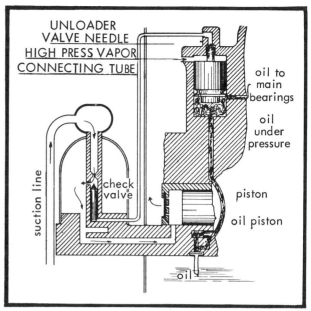

Figure 83

The sequence of operation is as follows:
When the unit stops running, the oil starts draining from under the piston that carries the unloader valve needle. This moves the needle off its seat, and high-pressure vapor passes through the orifice of the unloader valve into the connecting tube. This high-pressure vapor is carried to a check valve which closes so that no vapor can pass into the suction line. The connecting tube also leads into the compressor intake at a point between the check valve and the cylinder so that some of the high-pressure vapor passes through this opening, and into the cylinder. The load is now equal on both sides of the piston so that little power is needed to start the compressor. When the unit starts, the oil pump begins to build up oil pressure to lubricate the main bearings of the compressor and also to fill the cylinder underneath the piston carrying the unloader valve. The unloader piston is lifted by this oil. When the piston reaches the top of the cylinder, the needle closes and stops the passage of high-pressure vapor into the connecting tube. As soon as the motor reaches the required speed, the compressor evacuates the vapor from the connecting tube. This causes the check valve to drop, and vapor from the suction line enters the compressor in the normal manner. Remember that the compressor is surrounded by high-pressure vapor in this unit.

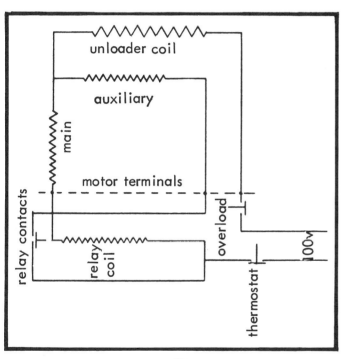

Figure 82

The General Electric Company and Servel, Inc. use an unloader that is operated by oil pressure. *Figure 83* illustrated the G.E. oil pressure unloader. The three parts that make up this unloader are: (1) the unloader valve and cylinder assembly; (2) the check valve, located in the suction intake of the compressor; (3) an oil pump to produce the oil pressure necessary to work the unloader valve. *Figure 83* shows the unloader valve and the check valve in the positions they should be when the motor is in operation.

SERVICING SEALED UNITS

At one time, if a serviceman cut into the refrigeration system for any reason, the five-year warranty was void. This generally restricted the repairs of these systems to refrigerators and freezers which were already beyond the five-year warranty and where the customer was usually unwilling to pay the very high cost of replacing the complete system, which could run to well over a hundred dollars. Furthermore, the manufacturer would not supply the component parts. The serviceman had great difficulty in obtaining exact replacement or general-purpose parts which would fit into the limited space readily. Service information was also difficult to obtain, which greatly handicapped the serviceman in the servicing of the more complicated units.

Now, the standard practice is to replace any component of a sealed system, such as the compressor or the evaporator, whenever replacement is required. The manufacturer not only sells the necessary replacement parts, but he supplies service manuals as well. In fact, there have been all out efforts made to aid the repairman to procure parts readily and at prices discounted so, finally, he can realize a profit.

No two models of refrigerators or freezers are exactly alike, not even if they are made by the same manufacturer. The manufacturer spends a great deal of time and money to design units that will provide the greatest amount of refrigeration at the least cost to him and in the most compact forms. Even the most experienced serviceman may sometimes have difficulty in determining the best way to get at a part which is to be replaced.

The following information will point out the generally accepted methods of replacing components, and point out some good alternate methods in special cases. Many times, however, a particular refrigerator or freezer is so designed that it requires a slightly different approach. While a good serviceman will eventually figure this out for himself, it may take a considerable amount of time and money for which· his customer may not be willing to pay.

The manual can often save him a considerable amount of time by pointing out short cuts which may not be evident at first glance. Most manufacturers are willing to supply recent service manuals to any serviceman, some free of charge, others at a nominal fee.

The next important statement we want to make is this: Do not be in too much of a hurry to replace parts of a sealed system. Replacing parts in a sealed unit is an expensive and time-consuming procedure and should be undertaken only after you have convinced yourself that the fault is really located in the system.

Some of the modern refrigerators are exceedingly complicated, and very often a failure in the electrical circuitry, in the controls, or even in the customer's way of using the controls may give rise to symptoms which are very similar to those you would expect from an inefficient compressor, a restricted capillary tube, or some other fault in the system. This cannot be emphasized too strongly: CHECK THE SYSTEM CAREFULLY BEFORE YOU CUT INTO IT. Therefore, before you cut into a refrigeration system, you should make the following checks: *Question the customer carefully as to the symptoms and then make the tests or observations which would be logical to make for these symptoms. Then, check all other possibilities and eliminate them one by one before you start to cut into the system.*

Most manufacturers keep very accurate records of the repairs made on their units, and they find that quite a large percentage of good compressors are being replaced. This means, of course, that the serviceman who replaced a compressor needlessly has not only cost the customer a considerable amount of money, but he has also probably not cured the fault. Obviously, you cannot get new repair business from such customers or from their friends.

For example, let us say that the complaint is excessive running time, accompanied by too high a temperature in the food compartment. Before you decide that the system is inefficient, check the door gasket for proper fit; check the amount of use the refrigerator is getting; check the actual temperatures in the food compartment and in the freezer or evaporator (very few customers have any idea of what these temperatures should be). Check the condition of the condenser-it may be badly clogged with dirt or lint. Finally, check the actual running time. If the refrigerator has become noisy, the user is often inclined to think that the running time has gone up because he or she is more conscious of its running. Sometimes, when a refrigerator has a short off-and-on cycle, it is thought to have long running time because the customer hears every time the unit starts, even though the running time may actually be normal.

After you have convinced yourself that the trouble is located in the refrigeration system, or in the electrical components associated with it, make the following tests to narrow it down further.

ELECTRICAL TEST SETS

Several test sets are available from manufacturer sources or they may be contrived by the serviceman. *Figure 84* shows one of the relay test sets available. It consists of a relay unit with two push-on type terminal connectors. It is a calibrated unit and should be treated like all good precision tools. The plug in relay is of the magnetic type and should always be operated in an upright position.

A manual test unit may be built of common electrical parts such as the one shown in *Figure 84*, or purchased complete at any good dealer supply house. Most dealers carry the component parts in stock to make a unit of your own. To connect the test units always identify the three terminals: *common (C), start (S), run (R)*.

To check out the circuit in the sealed unit follow the steps below:

(1) First pull out the fused receptacle from the test set (use a ten ampere fuse in this receptacle) and connect the two leads marked capacitor together so that they make a good contact.

(2) Connect the three marked test leads to the corresponding motor terminals. Plug the tester service cord into a live wall receptacle of the proper voltage.

(3) Hold the manual starting switch in the starting position with the thumb and insert the fuse receptacle.

(4) Within the period of three or four seconds release the starting switch. If the compressor is all right it will start and run in this length of time. It is not advisable to hold the starting switch closed for more than a few seconds or it is possible to burn out the starter winding.

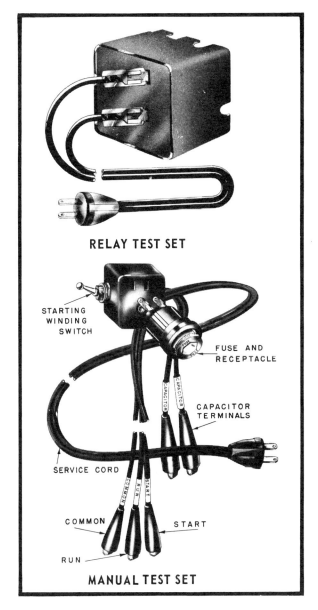

RELAY TEST SET

STARTING WINDING SWITCH

FUSE AND RECEPTACLE

CAPACITOR TERMINALS

SERVICE CORD

COMMON

START

RUN

MANUAL TEST SET

Figure 84

(5) If the compressor starts and runs with the test set, replace the relay.

(6) If the compressor does not start and run, disconnect the test set at once to avoid burning out the main motor because it is always connected in the circuit when the set is connected. If the compressor does not start and run with the set or the regular electrical accessories, check the line voltage. There should not be more than a 10% variation from the normal 115 volts. If the voltage is correct, replacement of the compressor is indicated. If the compressor tests good, the balance of the electrical circuit will have to be checked with the proper electrical schematic as

a guide. Be sure that all connections are good and clean. Look for broken wires, frayed insulation and loose terminals.

Important Notice: The use of a good troubleshooting manual with the proper specifications of the unit will go a long way towards simplifying the above process of compressor checking for electrical trouble, as well as for refrigerant information.

A very valuable book containing this needed information is "Tech-Master" published by Master Publications and available at most parts distributors. This book lists the start and run wattages, operating pressure, refrigerant and oil charges, thermostat operating temperatures, etc., for all domestic refrigerators and freezers. A copy of this book in your tool box will save you many hours of wasted time when diagnosing or repairing a sealed refrigerator system. Much of this information cannot be obtained from the manufacturer's literature.

TESTS TO MAKE IF THE COMPRESSOR DOES NOT RUN:

If the refrigerator or freezer has a force-draft condenser, see if the fan is running. If it is, the temperature control and the electrical system up to the unit itself are working properly. If the fan does not run, either the trouble is located in the temperature control or in the wiring system inside the cabinet. Check these points and correct the fault. If the unit does not have a forced-draft condenser, disconnect the refrigerator from the wall outlet. In some cases, you will find the compresser fitted with a short power cord of its own which leads to a terminal block; in other cases, the compressor has its own terminal block. By means of an extension cord, connect the 115—volt supply directly to the compressor, but make sure that the starting relay and the overload protector are still in the circuit. If the compressor starts, the trouble is obviously located in the temperature control or in the wiring inside the cabinet. If the compressor does not start, the trouble may be located in either the overload protector or in the starting relay, which should now be tested.

TESTING THE OVERLOAD PROTECTOR

The overload protector is generally mounted inside the terminal cover if the relay is separate from the overload protector. However, many refrigerator manufacturers use a "hot wire" relay, a combination of a relay and overload protector.

When the overload protector cycles, it is a normal function. The overload protector, as the name implies, activates when the compressor is overloaded. The problem could be high pressures within the compressor due to a component malfunction; turning the refrigerator control to higher setting just after a cycle and not giving the refrigerator time to equalize the pressure within the system; or failure of the compressor itself. Broken head or check valves within the compressor, binding bearings within the compressor, or moisture and subsequent corrosion, could all cause the overload protector to cycle.

1. Disconnect the power source.
2. Remove the overload from under the terminal cover.
3. With an insulated wire, jump across the two terminals, see *Figure 85*
4. Reconnect to power source. Turn on control if you have previously turned the refrigerator off.
5. IF THE COMPLAINT WAS FAILURE OF THE COMPRESSOR TO RUN OR HUM and it runs with the bypassed protector, then the protector device will have to be replaced. IF THE COMPRESSOR HUMS AND DOES NOT START refer to text RUNNING THE COMPRESSOR DIRECT WITH A TEST CORD, but because you do hear a humming sound, the protective device is not at fault. Check the relay and the compressor.

RELAY CHECK

If the compressor does not start, the overload protector is not at fault, and you should then test the starting relay. Examine the relay carefully until you locate the two terminals or wires leading directly to the starting contacts. Then, with your insulated screwdriver, or a short piece of insulated wire, short across the starting contacts. If the compressor starts, the relay is defective. *DO NOT MAINTAIN THIS SHORT CIRCUIT FOR MORE THAN A FEW SECONDS OR YOU WILL BURN OUT THE COMPRESSOR MOTOR.*

Figure 86 is a typical relay and overload protector assembly used in a refrigerator. The small terminal block labeled *S, C,* and *R* is the compressor terminal block: *S* is the starting winding, *C* is the common terminal, and *R* is the running winding.

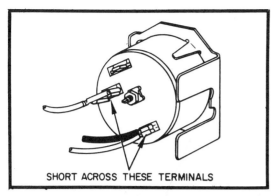

SHORT ACROSS THESE TERMINALS

Figure 85

TESTS TO MAKE IF THE COMPRESSOR IS RUNNING:

If the preliminary tests show that the trouble is actually located in the refrigeration system, the next thing to determine is in what part of the system the fault lies. Many a compressor has been replaced when the trouble was a restricted capillary tube or a leak in the system. For this reason, you should carefully check the entire refrigeration system before starting any replacement procedure.

If the compressor runs the normal amount, or all the time, yet, little or no refrigeration takes place, check the system as follows:

LOOK–LISTEN–FEEL.

Open the refrigerator door and listen for the usual gurgling or hissing sound of the refrigerant in the evaporator. Incidentally, if the refrigerator has a forced-draft condenser, be sure the compressor is actually running while you are listening to the evaporator. In some refrigerators or freezers with forced-draft condensers, the fan produces more vibration and noise than the compressor, and it is very easy to think the compressor is running when only the fan is operating. To make certain, put your hand on the compressor itself, or block the fan blades for an instant until you have made sure that the compressor is running. Then, free the fan blades so that the compressor will not overheat and cut off at the overload.

Another point to remember is that in any capillary tube system, when you shut off the system and switch it right back on again, the compressor will not start - the starting torque of the motor is not enough to start the compressor until the pressure in the system has had time to become balanced. As a result, the overload relay will cut out. A minute or so later, when the motor winding has cooled, it will cut back in. It will cycle on the overload this way without actually starting for three to five minutes until the system is balanced, after which the compressor will start and continue to run. When making tests which result in repeated starting and stopping of the compressor, allow time for the system to become balanced after each stop, or you may arrive at the wrong conclusion.

Now, if you do not hear any hissing or gurgling sound at the evaporator, there is no refrigerant flow, and one of three faults is probably the cause:

   a. A moisture freeze-up in the system.
   b. A refrigerant leak.
   c. A restricted capillary tube.

Stop the compressor and apply heat (a lighted match works very well) to the end of the capillary tube where it enters the large tubing of the evaporated. If there is sufficient moisture in the system to freeze up, this is where the freeze-up will occur, and because of the small diameter of the capillary tube, even a very small piece of ice will block it. Moisture can get into a system with the air drawn in through small leaks in the joints of the tubing in the low-side, at the compressor terminals, or even through small holes in an aluminum evaporator.

If the moisture in the system has frozen up at the end of the capillary tube, heating will melt it and you will hear a sharp hiss as the flow of refrigerant is resumed. If this occurs, you may be sure you have moisture in the system, but you can check by letting the refrigerator run a while. The compressor will pull down the temperature and operate normally.

The permanent remedy for freeze-up caused by moisture is to purge out all the old, moisture-laden refrigerant, add a drier, and recharge the system with fresh, dry refrigerant. The proper procedure for this operation will be described later. If you do not replace the wet refrigerant, the moisture will only freeze up again at the capillary tube and you will have another service call to make.

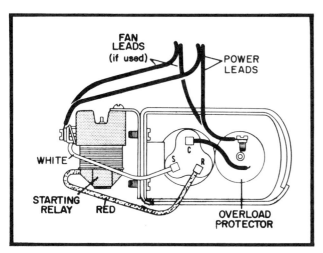

Figure 86

MOTOR CAPACITORS

The skilled refrigerator serviceman must be trained and possess much more technical knowledge than most tradesmen. He must be an electrician and a plumber. He must have knowledge of chemistry, he must be a welder. He must know instruments and have the knowledge to read test equipment.

He must have a vast amount of complex equipment, a volt ohmmeter, gauges, evacuation pump, charging cylinders, thermometers, temperature recording instruments and many more too numerous to mention. Besides all of these, he must have a well stocked tool box for hand tools, including flare tool, tubing cutter, hermetic kit, tubing benders, swaging tools and pinch off tools and perhaps many more. He must be proficient in all of the trades mentioned and especially so as an electrician.

As an electrician the skilled serviceman must know electricity and its application to motors. In recent years the capacitor motor, *Figure 87*, so called because it requires a capacitor in its circuit to start the motor, has been widely used. In air conditioning, in commercial refrigeration and domestic refrigeration, the capacitor type motor is most prominent. In domestic refrigeration the compressor starts with a capacitor. In air conditioning and commercial refrigeration a capacitor is used for motor or compressor starting purposes. Starting a capacitor type motor involves a capacitor, a centrifugal switch or a relay. The centrifugal switch cuts off the starting phase by using a mechanical means. When the motor reaches running speed or a little before, the centrifugal force of the motor causes the weighted mechanism to move a spool type device mounted on the motor shaft. This spool like object is usually made of an insulated material. It presses against a switch and disconnects the starting winding from the circuit.

Figure 87

The relay does basically the same thing except it is either voltage activated or current activated. See the section on RELAYS. In the early days of refrigeration, the repulsion induction motor was widely used. The repulsion induction motor was the real work horse for the refrigeration industry. It was made up of many components besides the windings and the armature. These included the carbon brushes, the carbon brush holders, the commutators, a commutator necklace and throw out mechanism, plus two steel pusher rods. Some of the early sealed units used in domestic refrigeration, such as the monitor top General Electric, the Majestic and Grunow, had capacitor motors. Eventually, as the trend went to sealed units, the repulsion induction motor was replaced by the hermetic unit which had the compressor and motor in a sealed case. The capacitor open type motor was used in many of the blower type of coils used in commercial refrigeration. They are also used in water towers for the cooling fan, and are used extensively on pumps, tools and equipment of all sorts.

TESTING THE CAPACITOR

Many refrigerators and freezers use a capacitor, *Figure 88* in the starting phase of the compressor. Often the capacitor will affect the starting of the compressor, and the unit will cycle on the overload protector continuously. If this condition prevails over a period of time, the compressor could be damaged. The best way to test a capacitor is to substitute with

Figure 88

another capacitor of the same or a higher rating. If a capacitor is not available for testing the compressor, proceed as follows:

CONSTRUCTING A TEST CORD

Figure 89 illustrates a test cord that can be used for testing the electrical components in refrigeration. Its uses can cover many other appliances, and care and pride should be taken in the construction, for it may become an important part of your equipment.

1. AC wall cap.
2. 14 gauge or heavier, 2 wire cord approx. 10 ft. long.
3. 3 alligator clips.

4. 2 lamp sockets.
5. 2 sockets to female plug adapters.
6. Single insulated wire 14 gauge or heavier, approx. 2 ft. long.

7. 1 momentary contact switch, 10 amps. or heavier. Normally open • N/O.

A test cord, similar to that shown in *Figure 90* may be used to test the system.

USING TEST CORD FOR CONTINUITY CHECK

Test with wires removed from component.
In using test cord as a continuity check, rig test cord as follows:

a. Insert 60 watt bulb in socket 1.
b. Use white and black leads only, insulate red lead.
c. Bulb will light on continuity.

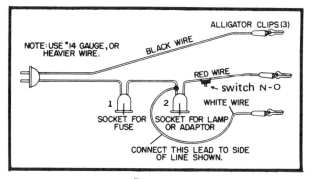

Figure 89

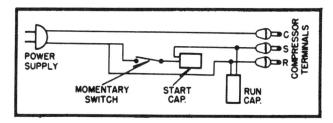

Figure 90 —*Constructing Test Cord*

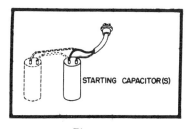

Figure 91

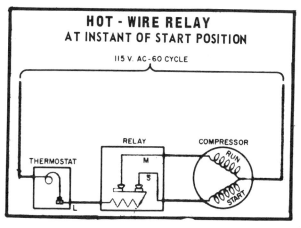

USING TEST CORD FOR CAPACITOR CHECK

Test with wires removed from components.

a. Insert 60 watt bulb in socket 1.

b. Use white and black leads only, insulate red lead.

c. Place leads across capacitor terminals.

d. If capacitor is good, light will glow slowly and dimly.

e. If capacitor is open, lamp will not light.

f. If capacitor is shorted, lamp will light promptly and brightly.

g. If line (d) tests as stated, replace bulb with 200 watt bulb.

h. Place leads across capacitor terminals, allow to remain a few seconds.

i. Remove leads, short across terminals with insulated wire, a visual spark should be seen if capacitor will load and discharge, indicating capacitor is good.

USING TEST CORD FOR COMPRESSOR CHECK

Remove all wires from compressor terminals.

a. Rig test cord for continuity check.

b. Test for winding burnt to ground internally, place one lead to clean tubing close to compressor, touch each of the three terminals, one at a time, with other lead. If bulb lights, compressor must be replaced and further testing would not be necessary.

c. If ground test is negative, proceed with further test as follows:

RUNNING COMPRESSOR DIRECT, USING TEST CORD

This test is made to determine if compressor will operate or if relay or other components are at fault. Test cord must be plugged into same wall receptacle or to the same line voltage. If possible it would be advisable to have an assistant help at the time cord is connected to wall receptacle.

a. Install 10 amp. fuse in socket 1.

b. Remove bulb from socket 2. , *Figure 91.*

c. Install adapter in socket 2.

d. Plug capacitor hook up into socket 2, *Figure 92.*

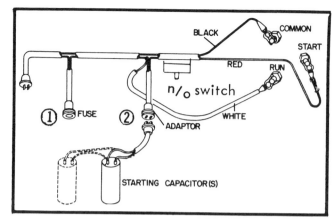

Figure 92

e. Remove insulation from red wire, all three wires will be used.

f. Connect leads as shown in *Figure 92*, white • run, red • start, black • common, refer to unit wiring diagram to identify compressor terminals.

g. Holding switch button down, connect test cord to service outlet or have assistant do this.

h. If compressor starts, release switch button, allow compressor to run for a few minutes.

i. If compressor fails to start and just hums, but does not blow the fuse, disconnect cord from wall, hook another capacitor in series with the existing capacitor, try test again.

j. If compressor fails to start, reversing red and white leads rapidly and with caution, while current is on and button is depressed, will often "bump" the compressor loose. If compressor fails to respond, remove test cord from wall receptacle.

k. Again try test, placing red and white leads back to the original position.

l. If compressor starts, release button, if compressor continues to run, allow it to do so, taking note if refrigeration is taking place. A restriction in the system could cause a compressor to "hang up"

m. If compressor starts, and runs after button is released on switch, but after a few moments stops, it could be from pressure build up as outlined in line (l), test would have to be made when system is discharged.

RESTRICTED SYSTEM

n. If fuse blows immediately on starting compressor, the compressor will have to be replaced. If compressor starts, but stops immediately after button on switch is released, compressor will have to be replaced.

Serviceman must bear in mind that an overcharge of refrigerant, or a restriction in the system, will cause a compressor to "bog down", so. before a compressor is condemned, make all tests possible to be sure. See text under *Figure 93*

Most restriction occurs at the condenser outlet (refrigerant strainer) or (capillary tube).

PARTIAL RESTRICTION—When a partial restriction occurs, the evaporator reacts almost in the same manner as with an undercharge. The suction line is warmer than normal, but the evaporator will likely frost in the area where liquid is present. If the restriction in the strainer or capillary tube is almost total, there is marked temperature difference in front of and behind the point of restriction.

COMPLETE RESTRICTION—When a complete restriction is present, both the evaporator and condenser are of room temperature and the wattage is below normal. This occurs when the evaporator has been pumped dry, the refrigerant is in liquid form in the condenser and no heat is being handled. With a complete restriction, once the compressor has stopped, it will rarely start again.

If a partial restriction is present, the condenser contains an above normal amount of liquid refrigerant and there is an above normal temperature difference between the top and bottom passes of condenser tubing. The discharge line will be extremely hot and the wattage lower than normal. It will take longer for the system to equalize when the compressor is stopped.

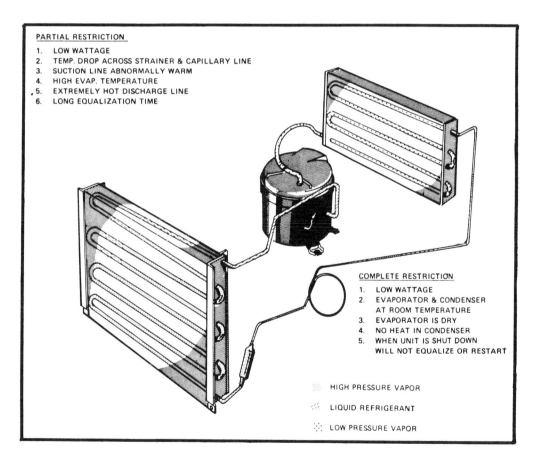

PARTIAL RESTRICTION
1. LOW WATTAGE
2. TEMP. DROP ACROSS STRAINER & CAPILLARY LINE
3. SUCTION LINE ABNORMALLY WARM
4. HIGH EVAP. TEMPERATURE
5. EXTREMELY HOT DISCHARGE LINE
6. LONG EQUALIZATION TIME

COMPLETE RESTRICTION
1. LOW WATTAGE
2. EVAPORATOR & CONDENSER AT ROOM TEMPERATURE
3. EVAPORATOR IS DRY
4. NO HEAT IN CONDENSER
5. WHEN UNIT IS SHUT DOWN WILL NOT EQUALIZE OR RESTART

HIGH PRESSURE VAPOR

LIQUID REFRIGERANT

LOW PRESSURE VAPOR

Figure 93   — *Restricted System*

HALIDE TORCH LEAK DETECTORS

Three similar detectors of gas leaks are shown in *Figure 94, 95 and 96* which differ only in the type of fuel used to furnish the flame; Prestolite, alcohol and propane.

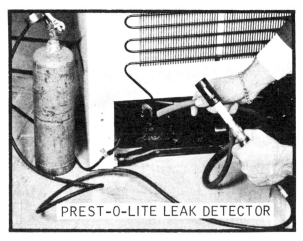

PREST-O-LITE LEAK DETECTOR

Figure 94

ALCOHOL LEAK DETECTOR

Figure 95

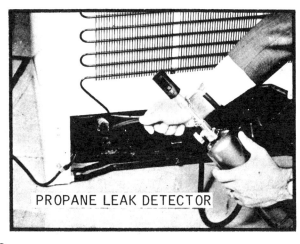

PROPANE LEAK DETECTOR

Figure 96

They are designed to detect leaks in the refrigeration system. Using the proper fuel with each detector, the following tests can be made: first, the small burner must be lit which will provide a blue burning flame. By the use of a tube, the system is progressively checked for leaks. *See Figure 97* Make sure that all fittings are also tested as well as any of the areas that were joined by solder in the original manufacturing process. Move the hose slowly over these areas and look for a change of color in the flame of the Hilide burner. A blue flame denotes no leaks, a light green tinge to the flame will show a small leak while the large leak will turn the flame either to a dark green or darker blue color. The refrigerant is non-combustible and so offers no danger when testing in this fashion.

HALOGEN LEAK DETECTOR

Figure 97

After considerable use the flame ring (reaction plate) in the tester may show oxidization or gather carbon. In this case, the reaction plate can be replaced easily by removing one screw and replacing it with a new one.

HALOGEN LEAK DETECTOR

A portable service tester has been developed by General Electric for testing Freon leaks and is called a Halogen Leak Detector. It can be pinpoint leaks ranging as small as 1/2 ounce per year which is a very small amount. When the leak is detected, an audible alarm will sound on the unit and continue as long as the probe encounters a leak. Instructions come on the unit for efficient application in all sorts of conditions and the unit operates on 120 volt 50—60 cycle power. *Figure 98* shows the probe being placed in position to make a test. If no signal is received, a special test unit is built in the tester to determine the setting required. This feature presets the test probe for all conditions and the results are very effective. The unit takes only a few minutes to "warm up" and then it is ready for use. Any leak will cause a loud growl to be generated in the loudspeaker.

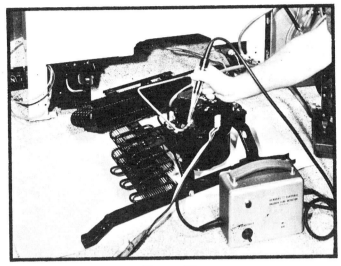

Figure 98

Note: there may be circumstances where it is necessary to use a soap solution for leak detection. If the cabinet is insulated with urethane foam, it is possible to get a false reading because *R–11* is the expanding agent for this insulation. The other methods are not reliable in this case.

USE LEAK DETECTOR

If the application of heat to the capillary tube does not result in even a temporary improvement, check next for a refrigerant leak with your regular leak detector. Leak test all highside tubing connections and joints while the compressor is running (this produces maximum pressure so that the leak is more easily detected), then shut off the unit and open the refrigerator door. Wait a few minutes for the pressure to build up in the low-side, then test carefully all the remaining joints in the system. If possible, warm up the evaporator, using ice cube trays filled with hot water, to build up the pressure. If the evaporator is made of aluminum, run the sampling hose on your leak detector very carefully along all the turns in the tubing to make sure you do not overlook any leaks. Important places to check are the buttweld joints where the aluminum evaporator tubing is joined to the copper tubing of the system. These joints are painted over, usually in an aluminum color, so you will have to look for them carefully or you may miss them.

If the system is provided with a forced-draft condenser, it is necessary to shut off the unit before any connections are leak tested.

If a leak is detected, either repair it or, if this is not practical, replace the leaky part. Instructions for the replacement of component parts in the principal system are given elsewhere.

If no leak is detected, check all tubing for sharp bends or kinks, paying particular attention to the capillary tubing. Sometimes, a sharp kink may cause only a partial restriction which will pass inspection, only to cause trouble eventually when the refrigerator is called on for maximum performance, as in hot weather and under full load. Usually, a kink can be straightened out well enough to be satisfactory- if not, replace the kinked tubing or the component of which the tubing is a part.

If these tests do not disclose the trouble, it is necessary to install a set of high - and low-side pressure gauges. This will enable you to determine whether the fault is located in the compressor or in some other parts of the system.

HOW TO INSTALL PRESSURE GAUGES IN SEALED SYSTEMS:

The following general instructions will apply to the installation of gauges in any standard domestic refrigerator or freezer. However, bear in mind that we cannot possible cover all the fine points or details of procedure which apply to every different model. For example, some sealed units are easily accessible from the back, so you need no particular instructions to get at any part required. Others are so compact and hard to get at, particularly the built-in models with forced-draft condensers, that you must pull the refrigerator away from the wall, lay it on its back or side, remove an access panel, and then pull out the complete condensing unit (compressor, condenser, fan, etc.) before you can do any work on it.

By far the best means of connecting a set of high- and low-side pressure gauges is through a gauge manifold, such as is shown in *Figure 99*. Such a manifold (also called a testing manifold) not only makes a much neater installation, keeping **the** two gauges together, properly aligned so that they can be read most conveniently, but it also permits you to charge, purge, cut into either the the high-or the low-side as required and perform many other service operations much more quickly and easily than would be possible with individual gauge connections. The gauge manifold not only enables you to do better work, but will pay for itself many times over.

Figure 99

For standard pressure checks, the gauge manifold should have three charging lines, one for each manifold connection. These may be made of 1/4-inch copper tubing, equipped with standard refrigeration type flare fittings, but we recommend the use of flexible hose with connectors at each end that can be tightened by hand. Copper tubing is stiff and unwieldly, and the flare connections will either break or leak after repeated use, requiring constant care. The connectors at the ends flexible hoses have internal neoprene gaskets which provide a good seal with only slight tightening of the connector nut.

HOW TO INSTALL A GAUGE MANIFOLD:

The exact procedure to follow in installing a gauge manifold on a sealed unit will depend upon the accessibility of the compressor and whether the refrigerator has to be laid on its back or can be worked on in a standing position. The following instructions will, therefore, cover the general principles - you should be able to figure out the details for yourself, or consult the manufacturer's service manual in case of a particular compact or difficult system.

There are three general methods of inserting the

testing manifold into the system. The method to use depends upon a number of factors, such as the accessibility of the compressor, whether you expect to replace the compressor or simple recharge the system, whether you prefer to use flare fittings, or if you are an expert in the use silver solder, and a number of other such considerations. These three general methods are:

    a. Using line tap valves.

    b. Using flare or other high-pressure fittings.

    c. Using silver solder fittings.

We will cover all three of these methods.

First, gain access to the compressor or to the high - and low-side lines leading to the compressor. You may have to remove cover plates and pull the unit away from the cabinet, as shown in *Figure 100* If the refrigerator is of the ultra-compact built-in type which has to be laid on its back or side, be sure to remove all food, loose dishes and shelves from the cabinet. If it contains frozen foods, these must be wrapped loosely in several layers of newspaper which will serve as temporary insulation until the refrigerator is back in service. If you must pull the compressor away from the cabinet, be very careful not to kink the tubing. Most refrigerators now use copper-clad steel tubing, rather than copper. This type of tubing is quite stiff and kinks very easily. Kinked tubing is not only difficult to stratghten again, but it becomes weaker and may in time crack or develop pin-hole leaks.

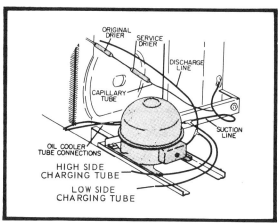

Figure 100

In some cases, the high-and low-side lines leading from the compressor are made of copper for a few inches so that the compressor can be more readily serviced. The copper tubing is welded to the steel tubing. Examine the compressor carefully, since all the tubing is usually painted black. You can tell where the copper tubing was inserted by noticing the weld at each end. The test manifold should be inserted in the copper tubing if there is any. Otherwise, select a point as close to the compressor as you reasonable can, but allow yourself enough working space so that you can use wrenches and a soldering torch. Usually, 6 to 10 inches from the compressor is about the right distance.

IMPORTANCE OF GAUGES

It is important that you have an accurate set of gauges that will register the true pressures in both the high side and the low side of the system, because it is these pressures that tell you much about the operation of the system.

The compound or "low side" gauge derives its name from the fact that it registers both vacuum and pressure. The compound gauge should be accurately adjusted so that the needle reads 0 lbs. when disconnected from the refrigeration system. It should also be checked occasionally for accuracy throughout its entire range.

Compound gauges that have been subjected to pressures above their range, or have become damaged through dropping or other misuse, may show accurate reading for short ranges on either side of the 0, but further up or down the scales they may be off enough to cause improper diagnosis of a system.

The high pressure gauge should also be checked from time to time to determine its accuracy and should be just as carefully treated as the compound gauge.

A check on the operation of a refrigerating system cannot accurately be made without the use of gauges. Sealed units, however, cannot always be checked with gauges. Therefore, you may have to resort to other means in checking these systems. If you watch the gauges carefully and learn how to recognize adnormal operating temperatures by feeling the various components while working on open systems, you will soon become familiar enough with the temperatures at the different points to enable you to diagnose troubles with reasonable accuracy in sealed units as well,

without the use of gauges. This method of diagnosing trouble will only come through experience, because you must have had the opportunity of checking many systems on which gauges could be installed in order to become familiary enough with these temperatures and to accurately understand the conditions that exist in a sealed unit.

INSTALLING LINE TAP VALVES

Line tap valves are quite satisfactory for emergency or temporary use, and for permanent use on the low-side of the system where the pressures are low. Regular fittings of the flare or silver solder type are safer for permanent use, particulary on the high-side, where the pressures may reach quite high levels in hot weather. A line tap valve consists of a body which fastens around the tubing to which it is attached. After it is securely fastened, a needle valve is turned to puncture the tubing wall. After the valve has been turned in as far as the directions call for, it is backed off again to permit pressure readings to be taken, refrigerant to be added, or the system to be purged, as required. The valve outlet must, of course, be securely capped, except for purging or when the valve is connected to a guage or charging line. When the service operation is completed, the needle valve is turned in again to close the punctured opening, and the outlet is capped. The valve is then left permanently in this position; it has an internal neoprene gasket around the needle valve to prevent leaks.

If you intend to use line tap valves, we recommend valves similar to the ones shown in *Figure 101* These valves come in a number of sizes, one for each tubing diameter, thus insuring a better fit than if they were intended for several tubing diameters. Line tap valves should be installed as follows:

Figure 101

Use a medium grade of steel wool or very fine emery cloth, clean a 3–inch section on both the high-side and the low-side lines at the locations you have selected.

Fasten the line tap valves to the tubing at the cleaned places, by following the manufacturers instructions carefully. Do not turn the valves yet.

Connect the testing manifold to the outlet ports of the line tap valves. Be sure that the low-side (suction side) of the manifold goes to the line tap on the suction line, and the high-side of the manifold to the high-side or discharge line of the compressor.

Connect a drum of refrigerant to the manifold. At this point, the manifold should look like the one in *Figure 102*. Be sure all connections at the manifold and to the refrigerant drum are tight, and that both manifold valves are closed (turned in to a firm seat). The connections to the line tap valve ports need be only tightened with the fingers at this point because later they will be used for purging the charging lines. Now, open the high-side valve of the manifold and loosen the charging line connector at the port on the high-side line tap.

Open the refrigerant drum momentarily and shut it off again quickly. The refrigerant will sweep through the high-side charging line and out the loosened connection, thus purging the high-side charging line of air. Give the line another short burst of refrigerant, then tighten the connection and close the high-side manifold valve. In the same way, purge the low-side charging line.

Turn in the needle valves on the line tap valves to puncture the tubing, and then back them out again. The gauges will now indicate the refrigerant pressure in the system. If the pressure is less than 40 pounds, open the refrigerant drum valve and let refrigerant enter the system until the pressure builds up to about 40 pounds, then shut it off again.

Test for leaks at all connections. You are now ready to test the refrigeration system.

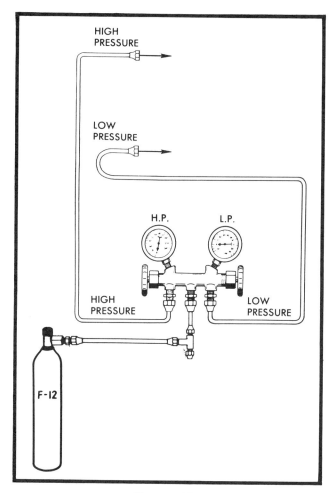

Figure 102

INSTALLING T FITTINGS

As a general rule, permanently installed T fittings are preferred over line tap valves in some instances. There is less chance of leaks developing, particularly in the high-side, and if you have to cut into the system to replace any parts, this offers a good opportunity to install these permanent fittings.

If you have the necessary equipment and skill with a soldering torch, you can use solder type fittings and silver solder. If you are unsure of your silver soldering technique, use regular fittings. Fittings are of two types: standard refrigeration flare fittings, and special high-pressure compression fittings. One of which is shown in *Figure 103* If the *T* fittings are to be inserted in copper tubing, flare fittings are the simplest, and often the least expensive. However, they are not suited for steel tubing because the fine seam in this tubing is very likely to split and leak when you try to flare it. The high-pressure compression fitting pictured in *Figure 103* has a neoprene gasket which conforms to any slight irregularity in the tubing and makes a tight seal. With either type of fitting, proceed as follows.

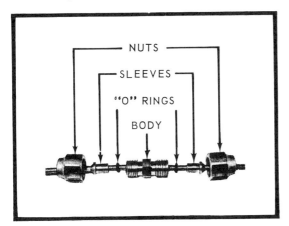

Figure 103

Using a medium grade of steel wool or very fine emery cloth, throughly clean a 2- to 3-inch portion of the suction line and of the discharge line and of the discharge line where you have decided to install the fittings.

Use a tube cutter to cut the suction line at the center of the cleaned portion. Work very slowly and carefully at first, making just a small opening for the refrigerant gas to escape through so that it does not shoot out too violently. Hold a cloth over the cut to catch any oil which may come out. When the hiss of escaping gas stops, finish the cut.

Cut the discharge line in a similar manner, except that you need not take precautions against rapidly escaping gas and oil. Insert the *T* fittings and either silver solder or tighten them securely, depending upon the type used. *Figure 104* shows the silver-soldered fittings installed as described here.

If you have chosen to use solder fittings, solder a short length of 1/4—inch copper tubing into the center port of the *T* and fit it with a regular flare nut so it may be connected to the testing manifold.

Connect the testing manifold, with its gauges, to the center ports of the *T* fittings you have installed. Be careful not to reverse the high-and low-sides when you make your connections.

Connect a drum of refrigerant to the center port of the manifold.

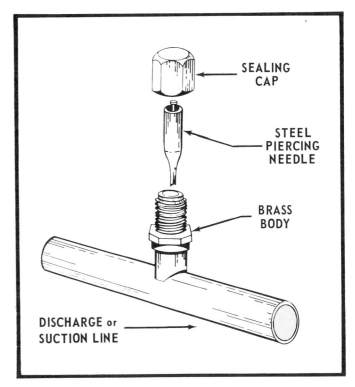

Figure 104

Open both the high-and low side valves of the manifold. Stand the refrigerant drum upright so that the charge will be added as a gas, not as a liquid, and open the valve on the drum.

If you are using a flare fittings, it is a good idea to purge the charging lines as explained in the preceding section for line tap valves. If the charging lines are soldered into the fittings it is not practical to purge them, but in that case you must be careful and do a thorough job of evacuating the system later when you recharge it.

Let the pressure build up to about 40 pounds, and then shut off the valve on the refrigerant drum.

Test for leaks at all connections.

The refrigerant drum can be left connected to the center port of the manifold for all the following tests. However, if it is in the way for subsequent tests it can be disconnected as follows. Close the valve on the refrigerant drum, if it is not already closed, shut off both manifold valves, and then remove the charging tube from the center port of the manifold. Cap the center port of the manifold to keep it clean and as a precaution against leaks.

PURGING, EVACUATING, AND CHARGING

When a sealed refrigeration system has to be cut into for any reason, a new drier must be installed and the entire system purged and evacuated before it is recharged.

To evacuate a system, you must have a good vacuum pump. Some servicemen dispense with the vacuum pump and use the compressor of the refrigerator itself for that purpose, but this is not universal practice. If the compressor is defective, it will not pull a sufficient vacuum. Even if the defective compressor is replaced before the system is purged, and then the replacement compressor is used to evacuate the system, you will contaminate it with any moisture or corrosion products which may have formed in the system because of the defect.

An exception to this practice is discussed in another section later. (See Gibson Refrigerators - Prior to 1966.

The actual procedure for purging, evacuating, and recharging a system is as follows:

Connect a gauge manifold to the high- and low-side of the compressor as explained in the section headed "How to Install a Gauge Manifold". Make sure that both valves on the manifold are closed.

Connect a line and charging T–fitting to the center port of the manifold, as shown in *Figure 102*

Connect a drum of refrigerant to the second outlet of the T–fitting again, as shown in *Figure 102*.

If you have followed the procedure outlined in "How to Install a Gauge Manifold", the system is charged with refrigerant which you have used for leak testing your connections. Discharge it by loosening flare nut "A" which connects the low-side to the manifold. Allow the refrigerant to escape until the high-pressure gauge reads zero.

Tighten flare nut "A" and open both high- and low-side valves at the gauge manifold.

SCRUBBING THE SYSTEM

Start the vacuum pump again. This double evacuation insures a better job than one of prolonged

evacuation, particulary with a service pump, which necessarily has only a limited pumping capacity. The second shot of refrigerant performs a sort of scrubbing action in cleaning out the residual air.

Connect a short length of copper tubing to the discharge port of the vacuum pump while the system is being pumped down, and when the compound gauge shows a vacuum of 25 inches, place the discharge tube of the vacuum pump into a glass containing refrigerant oil. You should see the discharged refrigerant bubbling through the oil.

Continue to run the pump at least ten minutes after all bubbling stops.

Close both manifold valves and stop the vacuum pump.

Open the valve on the refrigerant drum momentarily and shut it off again quickly. This will break the vacuum in the charging line.

REMOVAL OF MOISTURE WITH HEAT LAMP

Removing moisture with a heat lamp is sometimes called for and the following hook up is suggested. When a wet system has resulted from a low-side leak, a heat lamp in series with the motor windings of the compressor during evacuation will aid in drying out the system.

A low-side leak is one that has allowed a large amount of air to be drawn into the system.

The use of the heat lamps will aid the moisture removal because the temperature of the compressor is raised at the same time.

The second heat lamp may be directed into the evaporator (or upon the evaporator plate) as shown in *Figure 105* The wiring setup is shown in *Figure 106* A 250 watt bulb is recommended in the process as the combination will allow about 30 volts to heat the run winding.

The compressor should be rocked back and forth during final evacuation to break up the surface tension of the coil so that moisture can escape, *Figure 107*

The system is now completely purged and evacuated. The next step is to add the correct refrigerant charge. This may be done in several ways, but the following is recommended by most refrigerator manufacturers.

Figure 105

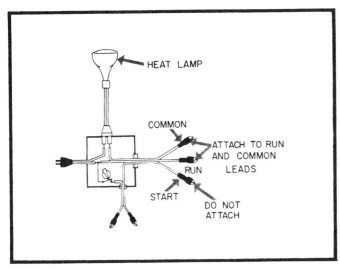

Figure 106

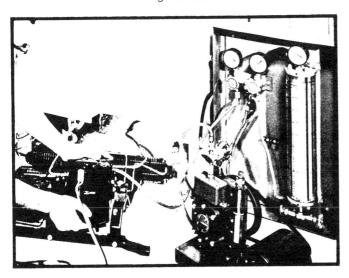

Figure 107

HOW TO ADD THE CORRECT REFRIGERANT CHARGE

The refrigerant charge in a modern domestic refrigerator or freezer is extremely critical. An overcharge or undercharge of one ounce may cause a serious difference in performance, particularly in two-temperature refrigerators. In such refrigerators the amount of refrigerant in the system is sometimes used to establish a balance between the operating temperatures of the main food compartment and those of the freezer. In one type of a two-temperature system, called the spillover system, the liquid refrigerant from the capillary tube flows into the evaporator of the main food compartment, where its evaporation lowers the food compartment temperature. Only one temperature control is used, and this is located in the main food compartment. The refrigerant charge is carefully measured for each model so that not all of it boils away in the evaporator of the main food compartment; some of it "spills over" as liquid into the evaporator of the freezing compartment. The entire design is carefully balanced so that the amount of refrigerant spilling over in this way is just enough to provide the correct range of temperatures in the freezing compartment when the main food compartment is maintained at the normal temperature range. To keep manufacturing costs down as much as possible, the accumulator at the outlet of the evaporator of the freezing compartment is quite small and has a very limited capacity. As a result, even a small undercharge will give rise to high temperatures in the freezing compartment; a small overcharge will cause liquid refrigerant to spill out of the accumulator, resulting in frostback of the suction line of the compressor.

This point is being emphasized to bring home the importance of adding a correctly measured charge, as well as the importance of checking the charge after it has been added.

Whenever possible, you should know in advance the correct charge for the sealed system being serviced. This information may be had from the manufacturer's service manuals or from their service departments. The refrigerant charge stamped on the serial number plate cannot always be depended upon. It is true that some manufacturers give the exact charge on the serial number plate, but others give the maximum charge for all systems using that particular type of high-side unit. The figure given on the plate can therefore be very much higher than is required for the refrigerator you are working on.

Generally speaking, when a serviceman goes out to repair a unit he knows he is likely to replace compressor, or at least cut into the system. In either case, he will have to add fresh refrigerant.

He should, therefore, find out in advance the correct refrigerant charge. He can then weigh out this exact charge into a small drum before he leaves the shop and use it only for charging purposes. There are refrigerant drums (sometimes called charging bombs) which hold up to 16 ounces of refrigerant. These drums are very handy for this purpose.

Such measured charges should be used only for charging. A regular 10–pound service drum should be carried for other service operations, such as purging the evacuating system.

Some servicemen prefer to carry their refrigerant in a calibrated charging cylinder, such as is shown in *Figure 108*. This cylinder can carry enough refrigerant for two or three jobs, and the serviceman can actually see the amount of charge being metered into system. If the charging cylinder is

used for direct charging, the refrigerant must be added as a liquid, since the cylinder cannot be heated. Generally speaking, it is better to add refrigerant in a gaseous state into the low-side, and this is the method we will describe here. The charging cylinder is then used only to put an accurately meassured charge into the charging drum.

Now, let us continue with the instructions for adding the correct charge:

If the exact charge is not known, or you do not have it accurately measured out, the service drum and vacuum pump should be left connected to the system. You can then add refrigerant as described in the following steps until the suction line frosts back to the compressor so you know you have an overcharge Then, purge a little at a time, checking the performance of the refrigerator each time, until you no longer get frostback and the operating temperatures within the refrigerator are correct. This, of course, consumes a lot of time and proves that it is important to know the correct charge in advance.

Open the low-side valve and start the compressor. With the charging drum in an upright position, open the valve and let the refrigerant gas (not liquid) enter the system. Heat the bottom of the

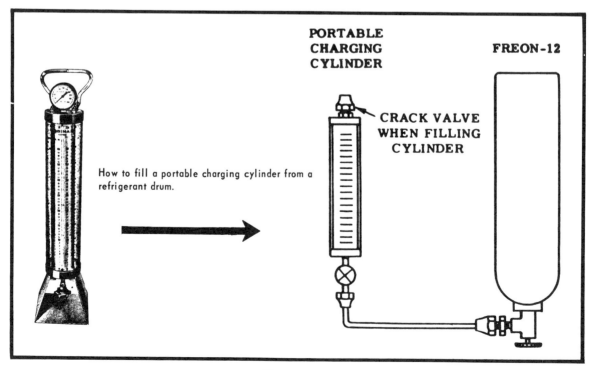

PORTABLE CHARGING CYLINDER

FREON-12

CRACK VALVE WHEN FILLING CYLINDER

How to fill a portable charging cylinder from a refrigerant drum.

Figure 108

drum slightly to aid in forcing the refrigerant charge into the system. Do not heat the top of the drum. It is common practice to incorporate a fusible plug in the valve body as a safety pressure release; such fusible plugs melt at approximately 165°F.

When the bottom of the charging drum remains warm after the application of heat has been discontinued and the compressor is running, you may be sure the drum is empty. Close the valve on the drum and also the low-side valve on the gauge manifold. Let the refrigerator operate through two or three cycles to be sure the charge is correct.

If the system has been overcharged, the suction line will frost between the evaporator and the compressor. If this happens, stop the compressor and purge off some refrigerant by loosening the low-side connection to the manifold momentarily and tightening it again quickly before too much refrigerant is lost. Wait until the frost has melted and then start the compressor again. Repeat purging if necessary, until no frostback of the suction line occurs.

If the system is undercharged, the temperatures in the refrigerator will be too high. The head pressure (high-side pressure) will be below normal, and the suction pressure (low-side pressure) will be somewhat below normal or just about normal, depending upon the degree of undercharge. If these symptoms are obtained, add more refrigerant. If too much refrigerant is added, frostback will occur, but this can easily be remedied by purging as described above. The correct charge is reached when the frost line appears approximately two to three inches on the suction line from the evaporator.

If the charge is correct and the refrigerator is operating normally, you are ready to seal off the charging lines that were connected to the center ports of the T fittings, or the line tap valves you inserted in the high- and low-side. This is done as follows:

If you have line tap valves, simple close off the valves, remove the charging lines, and cap the ports firmly.

If you have used solder or flare fittings, pinch off each charging line and seal it as follows:

Pinch off the tubing about 6 to 8 inches from the T fitting, using a standard pinch-off tool, as shown in *Figure 109*.

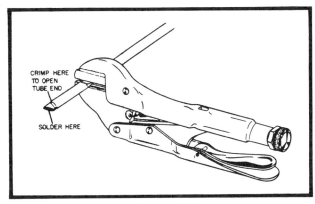

CRIMP HERE
TO OPEN
TUBE END

SOLDER HERE

Figure 109

Make a second pinch-off about 2 inches closer to the T fitting. Leave the pinch-off tool on the second pinch-off. The tubing usually springs open just enough to cause a leak unless pressure is maintained with the pinch-off tool.

Using a pair of side-cutters or side-cutting pliers, cut the tubing at the outside end of the first pinch-off so that a small opening is left into which solder can flow. If you have cut too close to the first pinch-off so that there is no opening for the solder to enter, crimp the tubing lightly with the side-cutters until the tubing just opens.

Solder the tube end closed, making sure that the solder flows smoothly into the opening for a perfect seal.

Remove the pinch-off tool and check carefully for leaks.

(GIBSON REFRIGERATORS - PRIOR TO 1966).

TO DISCHARGE, EVACUATE AND RECHARGE THE SYSTEM

1. Cut the Hi-side tube 4 inches from the compressor. NOTE: Only pierce the tubing to let the refrigerant charge out.

2. Install ¼'' flare nuts on each end of the tube cut in step 1.

3. Install a ¼ x ¼ SAE Union in the flare nut on the compressor hi-side.

4. Install a flare nut plug in the opposite flare nut.

5. Install a line-tap (piercing) valve on the compressor charging stub. See *Figure 110*

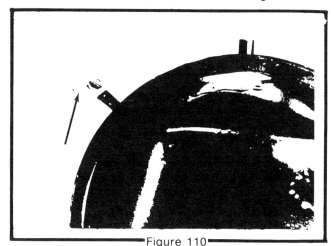
Figure 110

6. Install a charging line on the piercing valve Open the charging line valve so that a vacuum can be drawn on the entire charging line as well as the system.

7. Plug the service cord into the power supply and let the compressor run until a 25" vacuum is obtained.

8. During the process of evacuation, warm the dome of the compressor and all connecting tubing with a Prest-O-Lite torch. NOTE: Never warm any system part warmer than is comfortable to touch.

   Suspend a 100 watt light bulb in the aluminum evaporator section on one of the wire shelves. The combination of heat vacuum will vaporize any moisture in the system.

9. Cap the ¼ x ¼ union installed in step 3 and remove service cord from power supply.

10. Open the Freon Drum and allow refrigerant to enter the system until pressure between drum and system stablize then close charging drum valve.

11. Allow refrigerant to remain in system 15 to 20 minutes.

12. Remove cap installed in step 9 and connect service cord to power supply and allow the motor compressor to pull a vacuum of 25 inches.

13. If the system is being recharged after a leak

has been repaired or to correct a moisture condition, repeat steps 10,11 and 12.

14. By sweeping the system with refrigerant twice and heating the system parts, all moisture will be removed.

15. After a 25 inch vacuum is reached the 3rd time, cap the ¼ x ¼ union installed in step 3 and remove service cord from the power supply.

16. Open the charging drum valve and allow a positive pressure of from 5 to 10 pounds tp build up through the entire system. Note after the 5 to 10 pounds pressure is reached on the charging line gauge, allow the system to set with the charging drum valve closed, for 2 to 3 minutes to allow the refrigerant to back up through the capillary tube.

17. Remove the flare cap from the ¼ x ¼ inch union and the flare plug from the ¼ inch flare nut and immediately attach the flare nut to the ¼ x ¼ inch union. The system is now ready to be charged. NOTE: A positive charge of 5 to 10 pounds in the system allows the hi-side line to be connected with no possibility of moisture or air entering the system.

18. With the charging drum in an upright position, open the charging line and drum valves. Connect the service cord to the power supply. With the motor compressor running, draw refrigerant gas from the top of the charging drum until the complete evaporator is frosted. Close the charging drum valve.

19. Allow the freezer to run through 1 cycle and check the frost on the suction tube. Remove the L.H. throat moulding to gain access to the suction tube. The suction tube comes from the accumulator mounted at the L.H. top of the liner, and goes down the L.H. frost side of the liner. The frost on the suction tube should extend to a point opposite the strike location at the time the motor compressor cycles off.

OVERCHARGE

If the frost on the suction tube comes down below the strike 6 inches or more, the freezer is overcharged and should be purged until the frost goes back up the suction tube to the strike location. NOTE: Only purge a small amount of refrigerant at one time and then recheck the charge.

UNDERCHARGE

If the evaporator starts to defrost near the location of the accumulator (upper L.H. side) on the off cycle, the unit is undercharged. Add refrigerant gas slowly until the suction line is frosted at the point opposite the strike, at the time the compressor stops.

(In 1966 the Gibson Company announced a change in the process just described and now, due to design changes, the normal evacuating process using a vacuum pump is recommended).

HOW TO REPLACE A COMPRESSOR

Before you decide to replace a compressor, read carefully the early sections of this subject in which we discuss the tests that should be made to prove that the fault is actually in the compressor, and not in some other part of the system. After you have decided that the compressor must be replaced, proceed as follows:

Before leaving the shop, uncrate the service compressor and check its condition carefully to make sure that it has not been damaged in shipment. It is usually supplied with the manufacturer's replacement instruction manual. Read this manual carefully, since it will give you a lot of detailed information on points of procedure for the particular model you are repairing which will save you much time and trouble, and which cannot be covered in a general treatise on domestic refrigerators and freezers. Even if you prefer to follow a general procedure that is somewhat different ways of preforming most complicated service operations, you should nevertheless understand the manufacturer's procedure thoroughly so you can modify it to suit your own way of working or to follow the general procedure discussed in this manual. Since no two manufacturers have exactly the same approach to servicing sealed systems, the serviceman generally adopts a procedure which he likes best and uses for all makes of refrigerator, modifying the manufacturer's instructions as needed to conform with his method of working. In doing so, however, he must still be guided largely by the manufacturer's service manual, since special designs or featurers of a refrigerator often make it necesfor him to take additional steps or change his procedure to avoid serious trouble.

Check the type of fittings you will require; sometimes they are supplied with the compressor, sometimes not. Be sure you know the correct refrigerant charge; check to see whether the compressor contains the regular oil charge, or whether it is shipped dry. See if the compressor contains a holding charge, or a full charge of refrigerant, or none at all. The general practice is for the manufacturer to supply the compressor with a full oil charge and only enough refrigerant to provide a positive pressure (this is called a holding charge), but there are always exceptions and it is best to be sure. See if the compressor is shipped with a small charging drum and if so, how much refrigerant it contains. Here, again, there is no commonly accepted practice and every manufacturer has his own preferences. Since it is standard practice to replace the drier or add a new one every time a compressor is replaced, some manufacturers include service driers with all service compressors; other do not.

There will be other points to check, and you will learn from your own experience what they are.

At this point you have made all the necessary checks and tests and you are ready to go ahead with the actual replacement procedure. The refrigerator has been prepared so that the unit is accessible and the replacement compressor is is placed within reach. If you want to make a final check as to the possibility of tubing restriction, incorrect charge or inefficient compressor, install your test manifold and gauges as outlined in "How to Install Pressure Gauges in Sealed System," and proceed with the tests as explained in the appropriate sections of this subject. Most of the procedures for these tests have to be performed, also, when a compressor is being replaced, so that they do not take too much additional time. However, if you have proved to yourself that the compressor is at fault, you can save some time by proceeding as follows:

Disconnect the refrigerator from the wall outlet and disconnect the compressor itself at its electrical terminals to the same extent as the replacement compressor is disconnected from its terminals.

Using medium fine steel wool or very fine emery cloth, thoroughly clean a 3—inch portion of the compressor suction and discharge lines fairly close to the compressor where you intend to cut it free.

Using your tube cutter, cut into the suction line slowly until the gas starts to escape. Hold a cloth over the cut to catch any oil which may spray out due to the high pressure of the refrigerant. After the pressure is released, finish the cut. Cut the discharge line in the same way.

If the defective compressor is to go back to the factory, it is a good idea to pinch closed the open tubing coming out of the compressor. The pinch-offs need not be gas tight; just so they prevent too much air and moisture from getting into the compressor and oil from leaking out during shipment.

Another method which is used by some servicemen in cutting our compressors is to use a pair of side-cutting pliers, as shown in *Figure 111*. This will pinch the tubing closed as it is being cut, so the open ends of the tubing at the compressor need not be pinched closed. However, the pinched ends of the tubing remaining with the refrigerator will then have to be cut off with a regular tube cutter an inch or so inside the pinch to provide a clean, square cut for making the connections to the new compressor.

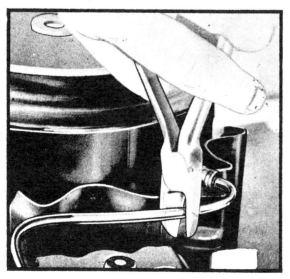

Figure 111

Lift out the old compressor and set the new one in its place. Connect the replacement compressor to the system, but insert T fittings and charging lines in the high- and low-sides as described in "How to Install Pressure Gauges in Sealed Systems". Be sure to add the drier also. If the manufacturer's manual accompanying the compressor gives specific instructions on drier installation, follow these instructions carefully. The type of drier to use and the preferred location for this drier can be very important with some refrigerators, and in such cases the manufacturer will usually provide the necessary information.

Connect a bulk drum of refrigerant (not the measured charge) to the center port of the manifold. Open both manifold valves and let the refrigerant enter the system until a pressure of at least 40 pounds is built up. Then, leak test the system. Close the valve on the refrigerant drum, and then close off both manifold valves. Disconnect the drum of refrigerant. Proceed from this point to step 3 of the section headed "Purging, Evacuating, and Charging".

After you have completed all tests and resealed the sealed system, mount the compressor in the refrigerator. Check the routing of all tubing to be sure there are no sharp bends or kinks. See that the refrigerant lines do not pass so close to each other or to other parts of the refrigerator that they will rattle when the unit is vibrating while running. If you have installed a heavy drier, be sure it is properly supported and securely fastened so that it will not vibrate and eventually cause the tubing to develop fatigue and crack.

HOW TO REPLACE THE EVAPORATOR, THE CONDENSER, AND OTHER MAJOR COMPONENTS IN A SEALED SYSTEM

There are so many variations in the design and location of the other major components of a sealed system, such as the evaporator, the condenser, the capillary tube or heat exchanger (this is the assembly consisting of the capillary and suction tubing which are either soldered together or are made of a double—tube of extruded aluminum to provide maximum heat transfer) that it is not practical to give more than general instructions for their replacement.

Read the manufacturer's service manual or replacement instructions very carefully until you thoroughly understand them. Then, follow these instructions to the letter if you can.

Modify them as little as possible to tie in with your own methods.

If no instructions are available, install a set of gauges and proceed from your general knowledge of refrigeration. Work carefully and slowly; think every step through before you do anything you cannot undo, and double check your work as you go along.

ALUMINUM CO-AXIAL SUCTION LINES

Aluminum co-axial suction lines may develop a leak on the low side. A kit is available from Whirlpool which eliminates the necessity of replacing the entire low side if a leak occurs.

The suction line is unbrazed from the compressor stub and the cap tube is cut at the drier. The heat exchanger can now be pulled out the front. Then the joint at the fitting where the capillary tube exits is unbrazed. By straightening the cap tube and cutting the suction line approximately 8 inches back from the free end, the section can be removed over the capillary tube.

The suction line can now be removed in sections by cutting it in convenient lengths for removal. Use the tubing cutter carefully so as not to damage the capillary tube.

The kit contains fittings to make the new installation with the full instructions with the kit will aid you in making a proper installation. There is a plastic sheath that will be installed to control electrolysis which probably caused the pitting in the first place. It's installation is covered fully in the instructions furnished by the manufacturer. To finish off the job the wet system is treated as explained in the "Removal of Moisture with Heat Lamp".

SERVICE PROBLEMS

Now that we have covered the designs and applications of sealed units, we want to take up the points to check when trouble develops. Many of the service problems in sealed units are the same as in open-type units. As an example of this, a temperature control functions the same way in a sealed unit as it does in an open-type unit, and it will break down in the same way and under the same conditions in both systems.

The temperature control and the wiring should be checked on any call where the complaint is long running time, too cold or too warm a cabinet, or unit won't run. Items to check are: (1) loose connections; (2) shorted wiring; (3) contacts burned; (4) contacts corroded so that no current flows; (5) power element may have lost its charge; (6) feeler bulb may be loose in clamp; (7) control may be out of adjustment.

The relay should be checked whenever the complaint is that the cabinet temperature is too high or the unit won't run. Items to check on magnetic relays are: (1) overload protective device, usually built into relay, because its contacts may be open: (2) partially shorted main winding may cause starting relay coil to hold starting contacts closed and trip overload device; (3) starting contacts may fuse together, causing overload device to trip; (4) relay will not operate if there is an open circuit in the main winding of motor or relay coil; (5) relay coil may be burned out.

Items to check on thermal or hot-wire relays: (1) main contacts may be open due to overload; (2) secondary contacts may not close because of a weakening of thermal wire; (3) starting contacts may have burned or fused together, causing overload to trip; (4) relay contacts may open too early, usually caused by tension of thermal wire changing so that less heating of the thermal wire is necessary to trip contacts open.

The motor should be checked whenever the compliant is that the unit will not run. The motor may fail to run because of an open or shorted starter winding, or it may be stalled because the compressor is stuck or overloaded as a result of low line voltage to the motor.

OPEN CIRCUIT IN MOTOR WINDINGS

Main Windings

Remove leads from the motor terminal and use a test light or ohmmeter on the motor terminals. If the bulb lights or continuity is read on any combination of two of the three terminals, the windings are not open. If the bulb fails to light when the test prods are on the main (M) and common (C) terminals, the main or run winding is open. If the start contacts of the relay are opened and closed manually, a small spark will be noticeable when the contacts are opened.

A thermal relay will trip open only the (S) contacts, and the motor will stop humming when these contacts open.

Secondary Winding

Remove leads from the terminals and use a test light in the same manner as outlined for main winding. If the light bulb does not burn when the test prods are placed on the secondary and common terminals, the winding is open.

A magnetic relay will close the starting contacts, but the motor only hums until the overload trips the circuit open. When starting contacts are opened by hand, there will be no sign of a spark at the contacts.

A thermal relay will trip the main contacts and the motor will have only one humming tone when current is flowing in the windings. Motor will not attempt to turn.

Another method of testing the motor is to use a relay known to be good and of the proper rating. Remove the motor leads at the motor terminals and connect the new relay directly to the motor and to a source of power. If the motor starts and runs properly, the trouble is not in the motor. If the motor fails to start and a magnetic relay is used, a spark will be observed when the starting contacts are opened. If a thermal relay is used and the motor does not start, the humming sound of the motor will change when the (S) contacts open.

SHORTED WINDING

A watt-meter should be used when a shorted coil is thought to be present in the motor windings

Compare the watt-meter reading with a chart furnished by the manufacturer of the unit. You will find that the wattage of the motor increases as the room temperature rises.

BROKEN LEAD INSIDE MOTOR HOUSING

If there is a broken lead from any of the terminals to the windings of the motor, the motor will not start. A broken lead to either the main or secondary winding will give the same indications as open windings in the motor. A broken lead from the common terminal to the windings is indicated by the motor being cool, and it will not hum with starting and overload contacts closed. Check for this condition by removing leads to the motor terminals. Place test prods of a test light on the terminals. If the bulb lights when the test prods are on the main and secondary terminals, but does not light when the test prods are on the common and either of the other terminals, the common lead is open inside motor housing. The same test can be made to show an open in the coil of a magnetic unloader on units where an underloader is used.

GROUNDED MOTOR WINDING

When a grounded coil is present, the sealed unit will usually fail to run. To test for a ground, remove the wall plug from the wall outlet, disconnect the leads from the motor terminals, plug in test light and place one test prod on bright metal of the motor housing.

Place the other test prod on each motor terminal in succession. If lamp lights when either of the terminals is touched, it indicates that the terminal is grounded. Reverse the plug in the wall socket and repeat the test. Lamp must light in both cases.

Practically all motors are grounded from the motor housing to the motor base. Some are grounded by means of a brass strip through the rubber mounting, while others are grounded by means of a separate ground lead. If the motor is grounded through one of the rubber mountings, it will be necessary to remove the motor from its base before testing. If the motor is grounded with a separate ground lead, this lead must be disconnected before testing.

LOW LINE VOLTAGE

A voltmeter should be used to check the voltage at the unit when it is starting. Low voltage will cause failure of the unit to start if it falls below 85 volts on any unit operating at 110 V. On many units the voltage must be kept higher than 85V in order for the relay to function. Low voltage will cause the unit to heat even though it runs.

MECHANICAL SERVICE

The mechanical services vary with the different types of sealed units.

Completely sealed units. Items to check when the complaint is long running time or the cabinet temperature is too high or too low:

> Door Gasket
> Location of cabinet in relation to hot-air registers and exposure to direct sun rays on condenser.
> Air circulation over condenser and dust and dirt on condenser.
> Excess frost on evaporator.
> Adnormal food or service load.
> Temperature control out of adjustment.

A temperature and time recorder will often aid in determining the trouble on sealed units where long running time is the chief trouble. Of course, cabinet repairs can be made on sealed units just the same as an open-type units.

Hermetically sealed units. In addition to the items listed for completely sealed units, you can check for:

> Air in the system.

> Take pressure readings on the high side of the system and determine if the refrigerant charge is normal or if the compressor is effecient.

> Purge or charge through the high side.

> Semisealed units; in addition to the things listed above, you can remove the motor-compressor unit from the fixture without removing any other part. The motor-compressor unit may be torn down and rebuilt in a service shop without the aid of special tools.

In recent years, many service firms throughout the country have equipped their shops with the necessary tools and equipment to completely service sealed units. Some firms even specialize in this kind of work. They are prepared to cut the unit open, rewind the motor, replace any parts, and weld or solder the housing back together. Upon completion, they usually guarantee the unit for 90 days.

FAILURE IN CONTROLS

There are, of course, certain failures in controls that must be taken into consideration. For example, the contact points on a control that make and break the electric circuit will frequently become burned or might possibly have been fused together from arcing so that they will not open again. Thermostatic controls may lose the pressure charge in the power element which will render them inoperative. When a power element loses its charge, it will always open the contact points, and, regardless of the temperature rise at the bulb, the contacts will not close again and complete the electric circuit. Also, the contact points may be closing properly but carbon deposits or other foreign matter on the points will not permit current to flow.

If a control is out of adjustment, it can be set to the proper cut-in and cut-out points by letting the condensing unit complete one operating cycle (one on and one off). A pressure control can be set by disconnecting the line from the control to the low side of the system and connecting this side of the control to a pressure-vacuum pump as illustrated in *Figure 112* The piston in this pump has a leather gasket on each side of the retainer washer so that when pressure is required, the gasket nearest the opening will prevent the air from seeping past. When vacuum is required, the gasket on the other side of the retainer washer comes into play and prevents air from entering the cylinder. You will note that a small hole is made in the tee in which the gauge and the half-union coupling are mounted. This hole is held tightly closed by pressing a finger over it during the pressure or vacuum test. By use of the pressure-vacuum pump, a pressure control can be correctly set in a few minutes without the necessity of waiting for the machine to go through a complete cycle.

Figure 112

Testing the electrical circuits requires the use of a test light that will indicate if power is being delivered to the motor. This instrument is a small pocket light, especially designed for checking electrical circuits. The test light will withstand 110 or 220 volts and, at the same time, indicate the voltage present in the lines.

A test light can be made by hooking two ordinary light bulbs in series. When 220-volt circuits are tested with this arrangement, the lights will burn brightly, while 110 volts will only make them glow.

In addition to the regular pressure and temperature controls, the auxiliary controls, such as those required to hold certain temperatures or pressures in the different parts of the system, may also need servicing from time to time. The main difficulty encountered in these controlling valves is improper seating due to foreign particles on the valve seat, or a worn valve gasket or needle.

All water-cooled condensing units should have a water valve on the water supply line to govern the flow of water through the condenser. If the water valve fails to function correctly, the condensing pressure may be extremely high, or else much lower than is needed. A water-cooled system should always be equipped with a high-pressure cutout incorporated into the pressure control used for starting and stopping the unit. This high-pressure cutout is a safety device which automatically stops the unit in the event the head pressure becomes too high. The high-pressure shutoff should be adjusted to stop the unit at a pressure of 140 pounds for sulphur dioxide, 175 pounds for methyl chloride, 180 pounds for Freon-12, and 300 pounds for Freon-22, on an average. This control should be hooked directly into the head of the compressor or into the discharge service shutoff valve port in such a manner that it is always open to the compressor pressure.

On water-cooled machines, the water valve itself may refuse to function because of foreign matter plugging the small orifice to the operating bellows, or a sticking valve stem. When the orifice in the connection at the top of the bellows becomes plugged, it may still have sufficient pressure in the bellows to hold the valve open. A more common occurence is that restriction in the orifice prevents the opening of the valve regardless of the operating head pressure of the machine. Insufficient refrigeration may be due to a restricted flow of water through the condenser as a result of improper operation of the water valve.

Failure of the water valve to open will cause the high-pressure cutout, installed on all water-cooled condensing unit, to open the electrical circuit and stop the machine. This high-pressure cutout is a necessary piece of equipment of water-cooled units because of the possible failure of the water supply and improper operation of the valve.

RELAYS

Sparks and arcing must be avoided in sealed units. This is accomplished by removing the centrifugal switch from the motor and replacing it with a relay. A relay may be defined as a device for opening or closing a local circuit under given conditions in the main circuit.

The relays in common usage for sealed units are the current-type (amperage), the voltage-type (potential), and the hot-wire relay. The current-and voltage-type relays operate by magnetism, and the hot-wire type operates by expansion and contraction caused by changes in temperature of the hot-wire element as current flow varies.

MAGNETIC ACTION

Since both the voltage and current type relays use magnetic action for operation, we will discuss them first. The magnetic relay makes use

of the fact that when an electric current flows in a conductor, magnetic lines of force are produced around the conductor. If the conductor is formed into a coil, the lines of force add together, and the strength of the magnetic field is determined by the number of turns of wire in the coil and the number of amperes flowing in the wire of the coil. In other words, a magnetic relay has an electro-magnetic coil as one of its parts. See *Figure 113*. The other parts of the magnetic relay are the movable armature which moves an arm that closes the contacts when the armature reaches the center of the magnetic field of the coil. An insulated mounting is necessary so that the electrical connections can be made to the relay. Another item usually found in a magnetic relay is an overload device. The relay provides a convenient place for mounting the overload, but it is not necessary to have the overload in the relay in order to make it function.

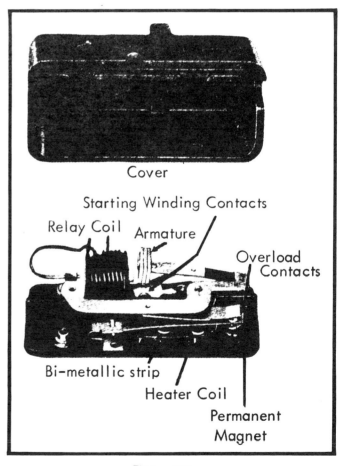

Cover

Starting Winding Contacts

Relay Coil   Armature

Overload Contacts

Bi-metallic strip

Heater Coil

Permanent Magnet

Figure 113

In current-type relays, the temperature control contacts, the overload contacts, the starting relay coil, and the main winding of the motor are all connected in series. The sequence of operation is as follows: The temperature control contacts close, completing an electrical circuit through the overload contacts, the starting relay coil, the main winding of the motor, and back to the side of the line. Electric current will flow through this complete path. Because the motor is not running, the number of amperes flowing will be large. This, in turn, will cause the starting relay coil (electro-magnet) to be strongly energized. The relay armature will be pulled into the center of the magnetic field, and, as a result, the starting contacts will close.

When the starting contacts close, a circuit is completed through the control contacts, the overload contacts the starting contacts, the secondary winding of the motor, and back to the other side of the line. We now have two magnetic fields present in the stator of the motor. These two magnetic fields act on the rotor and cause it to revolve. As the rotor gains in speed, the impedance of the stator windings increases and causes a decrease of current flow in both windings of the motor. When the current flow in the main winding decreases below a predetermined point, the strength of the magnetic field in the starting relay coil has also decreased and is no longer able to overcome the spring tension. This results in the relay armature rising, breaking the starting contacts, and disconnecting the secondary winding from the source of power. The motor continues to run as an induction motor until the temperature control opens the main contacts.

This covers the construction and operation of the modern, split-phase motor. The older style of split-phase motor has, in addition, a resistor between the starting contacts and the secondary winding of the motor. The capacitor-start motor has a capacitor instead of the resistor just described. In the two-value capacitor motor, the only other addition is a set of points on the starting arm. One set is known as the starting contacts and the other set as the running contacts. In a dual capacitor motor, the second set of contacts is not necessary because the running capacitor is in the circuit all the time the motor is running and is connected in paralled with the starting contacts.

HOT-WIRE RELAY

Another type of relay, known as the thermal relay,

or hot-wire relay, is manufactured by Delco. This relay operates on the principle which states that when an electric current passes through a conductor, heat is produced. *Figure 114* shows the construction of a thermal or hot-wire relay.

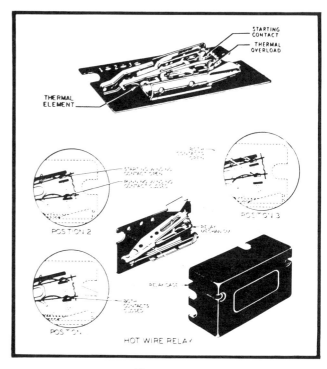

Figure 114

The wire controlling the contacts must be heavy and made of a material that has great resistance to the flow of current because we want it strong enough to withstand stresses while hot. A wire of small diameter would be more likely to separate.

There are two sets of connections on the back, labeled (S) for secondary or starting winding, and (M) for main winding. The contacts are closed by the force of contraction of the thermal wire, overcoming spring tension. When current flows through the thermal wire, the wire is heated and elongates. As a result, the (S) contacts are opened because the spring tension is then strong enough to move them. These contacts remain open during the running period of the motor because the current flow to the main winding of the motor keeps the wire warm and in balance with the spring.

A separate overload device does not have to be built into this relay because we already have a wire that is sensitive to current changes. This heat-sensitive wire can be used as an overload protector by simply making it break the main circuit whenever too much current is flowing in the main winding. This is the reason for the contacts labeled (M).

Care must be taken when installing a thermal relay so that the relay is of proper size for the unit; otherwise, damage will result to the motor windings. This is especially true if the relay is oversize, because then the starting contacts may close while the motor is in operation and the secondary winding may be burned out.

Some of the most common wiring diagrams for various types of relays and their connections are shown in *Figures 115* through *124* These do not cover all the wiring arrangements, but will be of great help for a number of those in common usage.

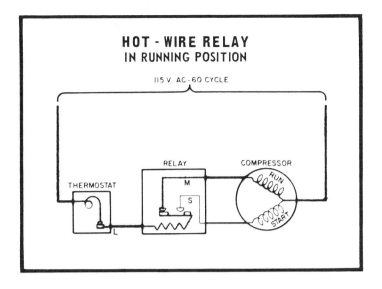

Figure 117

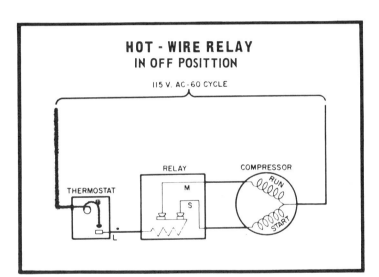

Figure 115

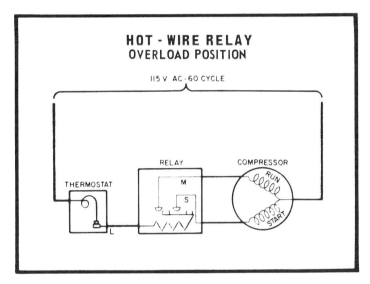

Figure 118

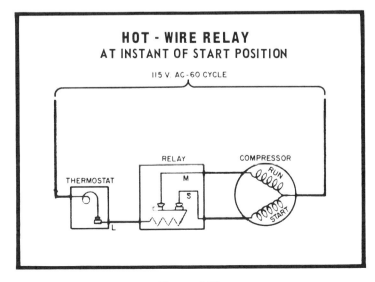

Figure 116

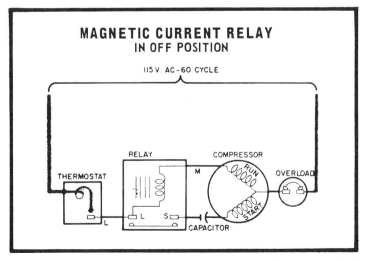

Figure 119

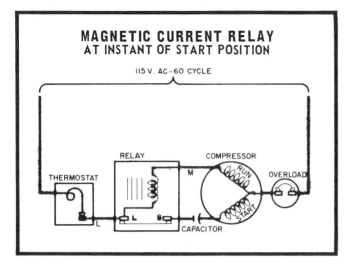

Figure 120

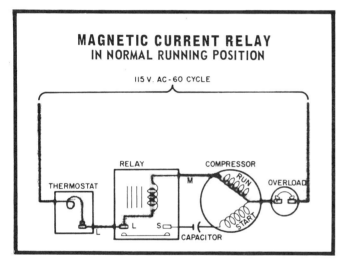

Figure 121

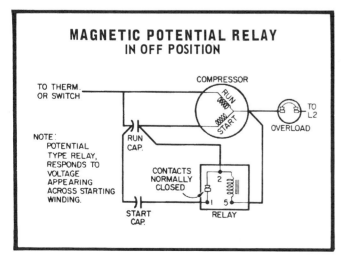

Figure 122

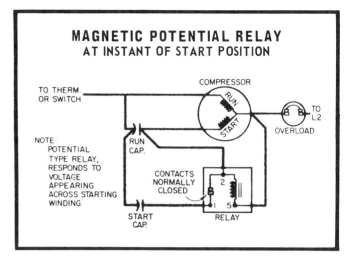

Figure 123

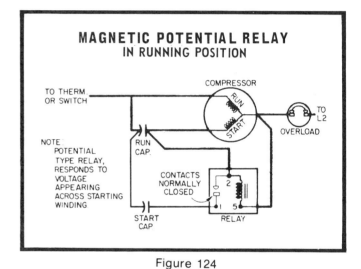

Figure 124

Shaded portions of electrical circuits in each diagram show the current flow.

In order to keep the temperature of the cabinet within definite limits, a control is used as on other machines to make the sealed unit operate part of the time and remain idle the remainder of the time. On domestic, sealed units, the most suitable control is the thermostat. The thermostat is not connected directly into any part of the system containing refrigerant. As a result, the control can be removed or replaced without difficulty.

Some manufacturers have special features built into their domestic units. These features include such things as: (1) secondary refrigeration systems, (2) special low-temperature compartments for frozen foods, and (3) refrigerant-cooled motors.

THE SOLID STATE RELAY

The Gemline Solid State Relay Part Number IC31, is available in kit form. The kit includes the hardware, terminals and accessories and will easily adapt to most refrigerators.

The solid state relay can be used on refrigerators that are rated from ½ thru ⅓ H.P. and easily installed. You must, however, use the old motor protector if it is still serviceable or purchase a replacement for it. The relay will replace thousands of styles and models presently in use. Because it has no moving parts, it will virtually last a lifetime. There are no points or movable contacts to pit and wear, and there is no magnet or magnet coil that could be troublesome. It is quiet in operation and is vibration and shock resistant. Only three wires are used with the relay, the wire fastened to the terminal marked L on the old relay would now be wire tapped using the wire nut. The other end of this same wire will fasten to the Run terminal on the compressor. If you do not know which is the run terminal on the compressor, follow the wire from the compressor to the terminal marked M on the old relay. The other wire from the IC 31 relay will then be placed on the Start terminal of the compressor, see diagram, *Figure 124*B .

The relay can be replaced with any type of relay as long as the rating is the same. A "hot wire" relay can replace a "current" relay; with the "hot wire" relay the protector device can be eliminated.

TESTING THE RELAY

Regardless of the type of relay, "hot wire" or the "current" type, testing the relay is not a complicated matter, but the reason for testing the relay could be. If the compressor starts intermittently, if it "hangs up" intermittently and cycles on the overload protector, the determination must be made whether the problem is in the refrigerator system, in the compressor, or if, indeed, the relay is at fault. There are three wires attached to the relay itself. On the "hot wire" relays, there are "dummy posts". These wires connected to the "dummy posts" are under the screws just to make wire connections and are in no way connected to the relay itself. So, we are concerned with the three wires to the relay proper. Look closely on the relay itself, *Figure 124*A. you will see the markings L for line, M for motor and S for start. Now that you have identified the wires, proceed as follows:

1. Remove the power source.
2. Remove the wires from the relay under the terminals L, M, and S.
3. Connect wires L and M together or fasten them under one screw on the relay for convenience.
4. Have an assistant plug into the power source.
5. Cautiously touch the S wire to the L and M wires.
6. If compressor starts, immediately remove the S wire. If compressor continues to run normally, the relay will have to be replaced.

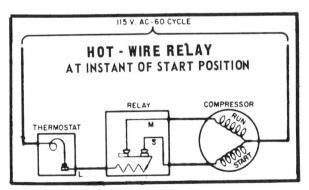

Figure 124 A

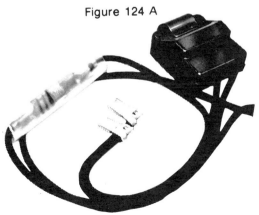

Figure 124 B   *Courtesy of Gem Products, Inc.*

SECONDARY REFRIGERATION SYSTEMS

Figure 125 shows a cutaway of a unit with one shelf refrigerated by means of secondary refrigeration system. You will note that the tubing of the secondary system is not connected into the main refrigerant line at any point. Instead, we find that the tubing is brought up to the secondary condenser and passes inside the shell of this secondary condenser. Usually, a full turn is all that is needed. The refrigerant in this closed path absorbs heat from the shelf and boils. The vapor rises, and when it comes into contact with the tubing clamped alongside the secondary condenser, it gives up that heat and condenses again. The heavier liquid falls to the shelf and the cycle is repeated. The secondary refrigerant continues to absorb heat because its vapor is continuously being cooled in the secondary condenser.

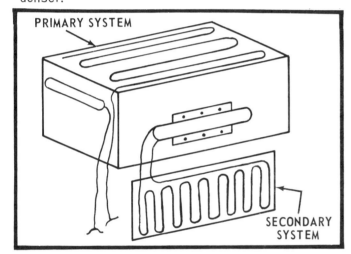

Figure 125

Care must be exercised in the use of these shelves because if the primary unit is operated at too low a temperature, this will increase the heat transfer of the secondary refrigerant and may result in freezing foods placed on the shelf.

The secondary refrigerant tubing may be placed around the inner liner in the insulation of the cabinet. The idea of using a secondary system is to keep the cabinet cool without frost formation and to maintain a higher humidity in the food area. If the walls sweat too much, a method of correcting this is to loosen the clamp holding the tubing of the primary liquid line to the secondary condenser and to place some material between the tubing and the secondary condenser. Care should be exercised when this is done so that too much insulation is not added, which would kill the effect of the secondary system entirely.

Figure 126 shows a system which is divided into two parts so that the lower space may be used to make ice cubes and store frozen foods. The upper space is kept at temperatures above freezing and is used to store fruits, milk, and other produce of like nature. Each space has its own evaporator. The upper space also has a sterile ray lamp to aid in the elimination of mold and other fungus growth.

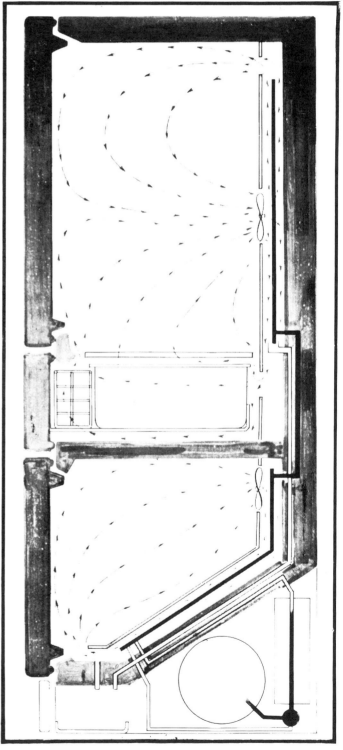

Figure 126

The evaporator in the colder compartment is kept at a lower temperature than the evaporator in the warmer compartment. This is accomplished by using a *Differential Pressure Control* (D.P.C.) between the two evaporators. This device is very similar to a loaded check valve. It has a spring acting against the needle valve, tending to hold the valve on its seat. The suction pressure in the colder evaporator must become low enough to overcome this spring pressure in order to open the valve. This allows liquid and vapor from the warmer evaporator to pass through. Thus, a pressure (and temperature) difference is maintained between the two evaporators. This valve operates with a throttling action and is not adjustable.

Sealed units have been used for several years in small, commerical freezers and coolers, especially in self-contained units. Sealed units have also proved to work quite well in ice-cream cabinets, beverage coolers, and packaged air conditioners.

Beverage coolers include beer coolers, water coolers and bottled goods coolers, sealed package air-conditioning units are built in sizes up to 5 hp.

The use of semisealed units as field replacements for open-type condensing units is increasing. These units are manufactured in sizes up to 100 hp.

TROUBLE SYMPTOMS

The symptoms which indicate that trouble is brewing in a refrigerating system are not too readily recognized by the customer who usually does not call the serviceman until the temperature in the refrigerator has become either too high or too low. Accurate and speedy diagnosis of the trouble is then required in order to get the system back into proper operation as soon as possible.

In order to correctly diagnose the symptoms, the serviceman must understand the function of each part in relation to all the other parts of the system. He must also understand in what way the improper operation of one part will affect the overall operation of the machine and be able to determine from these symptoms what part or parts, or conditions, are causing the trouble. Many of the troubles in a refrigeration system may be the result of a complication of existing conditions.

It is not always possible to immediately determine the cause of improper operation, because it is sometimes necessary to first eliminate several of the possible causes by a systematic check of the function of each part.

When you undertake to service a refrigeration system, it is necessary that you have the proper tools and instruments to carry out your work effectively. Very often this includes the correlation of information obtained from the thermometer and the gauges, together with the tell tale noises that a machine out of order will make while in operation.

DIAGNOSING VALVE TROUBLES

When diagnosing troubles in a valve, first consider all the conditions that may prevent the valve from working.

A low-side valve may be mechanically perfect, but a plugged screen, a low refrigerant charge, or improper positioning or the entire low side may prevent the valve from doing its job properly. A high-side valve may cease to function correctly as a result of an improper refrigerant charge, the valve not being level, or some other condition which may prevent the liquid refrigerant from entering the valve chamber. If all the other parts function correctly and the valve is actually the part which is causing the difficulty, ascertain whether it is leaking or incorrectly calibrated.

These same conditions effect the operations of both thermostatic and automatic expansion valves, even though the valves themselves are in perfect condition.

The thermostatic expansion valve is probably the most misunderstood part of the refrigerating system. This is probably because the function of the thermostatic expansion valve is often confused with that of the automatic expansion valve.

The automatic expansion valve is pressure controlled, whereas the thermostatic expansion valve is temperature controlled. The function of the thermostatic expansion valve is to keep a sufficient amount of refrigerant in the evaporator to maintain the temperature at the predetermined setting regardless of load changes. Always keep this in mind when servicing a system. Adjustment of the T.E.V. merely floods or starves the evaporator by changing the superheat setting. A thermostatic expansion valve seldom changes adjustments of its own accord. This means that if the valve has been operating satisfactorily and suddenly changes, the chances are that the adjustment has not been changed, but something else has gone wrong. The other parts of the system should therefore be checked before attempting to adjust the T.E.V. If you wish to check whether or not the T.E.V. will open, this can be done by removing the bulb from the suction line and warming it by holding the bulb tightly in the hand.

One of the most common causes for expansion valve failure is water in the refrigerating system. Minute quantities of water in the system will eventually collect at the expansion valve orifice and freeze the needle into its seat. It does not require more than a small drop of water to stop the operation because ice at this point will act as though the needle were soldered to its seats. After the expansion valve warms up and the ice melts, the valve may again open, but the first passage of liquid refrigerant will immediately freeze the water with the same results. Enough of the water may be carried along with the first rush of refrigerant through the orifice to free the valve for some time. However, the water will collect again at the valve orifice and cause the same trouble.

Automatic expansion valves are also subject to freezing in that moisture condenses and then freezes within the adjustable spring chamber of the bellows. When water freezes in the bellows, it will restrict the freedom of movement of the bellows and may either hold the valve wide open or tightly seated, depending on the position of the needle at the time of freezing. Freezing in the bellows of a T.E.V. is also not impossible, but is not as common as freezing in the bellows of an automatic expansion valve.

It is not all uncommon for a system to operate for as long as a year without trouble, and then suddenly to freeze up. The moisture can very easily remain in the system (either in the evaporator or in the crankcase) for long periods of time and finally work itself out and start circulating along with the refrigerant.

RESTRICTION IN THE SYSTEM — INEFFICIENT COMPRESSOR — REFRIGERANT SHORTAGE

These three defects have many symptoms in common; the net result is the same - insufficient refrigeration. The proper approach in determining the exact defect is described elsewhere so that this procedure should be followed up to and including that part which describes the testing of the compressor for pumping ability. If the compressor tests OK and no leaks exist, purge, evacuate and recharge the system until it contains the correct charge. Start the compressor and watch the high- and low-side pressures after two or three operating cycles. If you have worked on this particular model before and know what pressures to expect after only one operating cycle, you can save some time, but otherwise it is best to let the evaporator get down to the operating temperature and take the pressure readings at the end of the second or third running cycle, just before the unit shuts off, with the temperature control at the normal position. Consult the manufacturer's service manual or the compressor replacement instructions for the correct operating pressures.

If the high-side pressures are higher than normal and the low-side pressures are normal, the capillary tube is restricted; such a restriction is usually caused by powdered dessicant from the drier or metal filings accumulated at the entrance to the capilary tube. As a check, shut off the unit after it has run long enough to build up to maximum head pressure and note how long it takes for the high- and low-side pressures to equalize. The normal time is about 6 to 10 minutes (this varies with the models). If it takes appreciable longer, the capillary tube is restricted and should be replaced. In that case, it is usually necessary to replace the heat exchanger (the capillary tube and its attached suction line) as a unit.

If the high-side pressures are normal or below normal, but the low-side tubing may be restricted. Look for kinks in the low-side tubing. The low-pressure side of such a restriction is always colder, so look carefully along the tubing for signs of frosting or sweating. If the evaporator is made of aluminum, look for an obstructed butt weld at the end of the aluminum tubing or for sloppy-looking joints in the aluminum tubing,
Figure 127

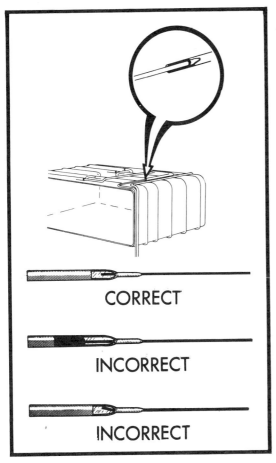

Figure 127

INCORRECT REFRIGERANT CHARGE:

An overcharge will result in frosting of the suction line between the evaporator and the compressor. An undercharge will result in insufficient refrigeration accompanied by lower than normal head pressures and somewhat lower than normal to normal suction pressures, depending upon the degree of undercharge. The procedure to be used in remedying these conditions is described under "Purging, Evacuating and Charging"

FAULTY SYSTEM PERFORMANCE

High head pressure in an air-cooled condenser due to improper cooling of the refrigerant will frequently be encountered in refrigeration service work, because air-cooled condensers are subject to reductions in air flow by lint, animal hair, and dust picked up by the fan and forced against the fins of the condenser.

When an evaporator is not performing correctly, it may be due to (1) improper feeding of the liquid refrigerant, (2) excessive load in the cooler, or

(3) inability of the evaporator to return the oil to the crankcase, thereby setting up an insulating blanket of oil that will prevent effective evaporation of the refrigerant.

It is advisable to use an accurate thermometer in testing the temperature of an evaporator in order to compare the evaporator temperature with the actual pressure in the low side of the system. The thermometer can be placed in the ice-cube tray compartment of domestic evaporators, preferable without removing the ice-cube trays. The thermometer should be slipped in between the trays and the side of the evaporator. The door of the refrigerator should then be closed for ten minutes or more to allow the thermometer to reach the lowest possible reading. Commerical evaporators are more exposed to air currents which will keep the temperature indicated by the thermometer from three to five degrees above the actual evaporator temperature.

A thermometer with a metal spring clamp attached in such a manner that the clamp can be slipped over one of the evaporator tubes will give a fairly accurate temperature reading. If more accuracy is desired, the thermometer may be insulated from the currents by insulating two or three inches of the tubing on each side of the thermometer bulb.

The thermometer test can also be used on condensers to determine the approximate temperature rise of the air through the condenser. If the room temperature is first checked, this will determine the ambient temperature. Then, by holding the thermometer in the air stream leaving the condenser, the temperature rise of the air can be closely estimated. A temperature rise of 15 to 20 degrees above the ambient room temperature is not uncommon. However, the size of the condenser and the characteristics of the machine, as well as the suction pressure at which it is operating, will reflect on the condenser temperature and on the pressure in the high side of the system. It is advisable to check a number of condensers under different operating conditions to get a general idea of how high the temperature of the air does rise in passing over the condenser surfaces.

UNIT CONTROLS (COLD CONTROL)

Control units can be mounted so that they "read"

the temperature changes directly from the refrigerated surface or from the ambient air surrounding it. The latter method has become popular in the late style refrigerators. It is particularly suitable for forced air systems. Being energized or deenergized by the air flow coming out of the food compartment, this method is particularly sensitive to actual area temperature changes due to food load requirements.

Cold controls or thermostats have developed over a period of time, and today there are many types and designs. One of the first types was a low pressure switch which operated by low side compressor pressure as its temperature reached a pre-determined point. A bellows arrangement contracted and stopped the motor by separating contact points. As the refrigerant absorbed more heat and the pressure rose, the action was reversed.

A basic thermostat design was developed which had a tubing of small diameter between the bellows and a spot on the evaporator with a closed end that clamped to the evaporator. Inside was a refrigerant gas that closely assumed the same temperature as the gas in the evaporator. The pressure, which varied much the same as the main gas charge changed with the temperature and activated the bellows switch. The vapor filled tube thermostat is now quite popular.

Improved action was obtained by working the bellows in conjunction with a magnet and an articulated arm carrying one of the points. In this manner a snap action was used to make the points open and close fast so that burning and pitting was minimized.

In *Figure 128* another type of snap action uses a toggle spring that delays movement until the arm has passed mid point of travel and then it moves rapidly. A dial on the outside of the control box controls the tension of the arm to delay the action as the dial is advanced. This delay causes the contacts to remain on longer, pro

ducing colder temperatures before the switch opens and stops the compressor.

Figure 128

An altitude adjustment is a part of the built in adjustment that can be adjusted for unusual locations. *See Figure 129* for the adjustment instructions for different make controls.

A thermostat differential is that difference in degrees that represents the cut-in temperature and cut-out temperature. It will vary with the make and model of the refrigerator.

AUTOMATIC DEFROSTING

Automatic defrosting is accomplished in several ways according to the individual design of the refrigerator. When the original design was developed there appeared to be more ways to do this than there were refrigerators. By now, it has settled down to two principle systems - hot gas defrosting and defrosting with electrically heated appliances in the system such as mullions and heat tubes. In most of the modern systems there is a combined use of these two methods. To speed the action it is often found necessary to have several heat tubes of different manufacture as well as the extra heater circuits known as dew point heaters. These dew point heaters are usually found around the doors to offset the sweating of the area that causes drips and moisture conditions around the pilasters.

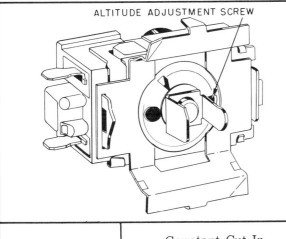

ALTITUDE ADJUSTMENT SCREW

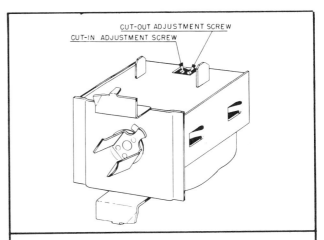

CUT-OUT ADJUSTMENT SCREW
CUT-IN ADJUSTMENT SCREW

	Constant Cut-In
Altitude Above Sea Level - Feet	Altitude Screw Adjustment (Turns Clockwise)
1000	No Change
2000	1/16
3000	1/8
4000	3/16
5000	1/4
6000	5/16
7000	3/8
8000	3/8
9000	-

	Variable Cut-In
Altitude Above Sea Level - Feet	Range Screw Adjustment (Turns Clockwise)
1000	3/32
2000	3/16
3000	7/32
4000	1/4
5000	3/8
6000	7/16
7000	15/32
8000	1/2
9000	9/16

ALTITUDE ADJUSTMENT

<u>Both</u> Cut-In And Cut-Out Screws <u>Must</u> Be Adjusted

	Constant Cut-In	
Altitude Above Sea Level - Feet	Turns Counter-Clockwise	
	Cut-In Screw	Cut-Out Screw
2000	1/8 CCW	1/16 CCW
3000	7/32	1/8
4000	5/16	5/32
5000	13/32	7/32
6000	1/2	1/4
7000	19/32	5/16
8000	11/16	3/8
9000	13/16	13/32
10000	15/16	7/16

	Variable Cut-In	
Altitude Above Sea Level - Feet	Turns Counter-Clockwise	
	Cut-In Screw	Cut-Out Screw
2000	1/16 CCW	1/16 CCW
3000	1/8	1/8
4000	5/32	5/32
5000	3/16	3/16
6000	1/4	1/4
7000	5/16	5/16
8000	3/8	3/8
9000	13/32	13/32
10000	7/16	7/16

Figure 129

The hot-gas method operates in direct reverse of the refrigerating cycle and requires compressor operation. Hot compressed gas leaving the compressor is by-passed around the condenser and the small diameter line into the evaporator coil of the freezer. The latent heat (and some sensible heat) contained in the hot gas is absorbed by the evaporator coil and condensed back into a liquid. The heat that is absorbed will melt the frost. From here it goes to the refrigerator section (sometimes called the food or provision compartment) and passes through the evaporator plate. The liquid and vapor refrigerant (mixed) is then drawn back into the suction line of the compressor dome. The hot compressor will vaporize the refrigerant some more. This vapor is drawn in by the compressor pump and the cycle repeats. See *Figure 130* for the defrosting cycle and *Figure 131* for the normal refrigeration cycle. It's action is easy to trace. There may be a slight difference in some of the systems but they are basically the same. The valve that bypasses the hot gas in the first place is called the defrost solenoid valve. Most of them are similar in design but may take on different forms to meet patent requirements. The valve can be of the normally-closed variety and opens when the defrost control energizes the solenoid.

In some types the suction pressure will rise in accordance with the evaporator plate temperature during the defrost period. The compressor dome temperature will drop to somewhere between 25° and 40°F above room temperature. In this case the wattage will increase. The best way to check the condition of the defrost system is to compare the specifications of the particular model at hand and follow the manufacturer's instructions. Not all types will give the same readings nor will they be controlled in the same manner.

DEFROST THERMOSTATS

A defrost thermostat will often have a constant cut-in temperature setting while the cut-out temperature will vary according to the dial setting. This is common with defrost models with evaporator plates.

The temperature at which the thermostat cuts-in is (approximately) 33°F. (just above freezing) This allows the evaporator to be defrosted on the "Off" cycle and is known as variable differential.

THE THERMAL BARRIER

A plastic sleeve, or covering is used by some manufacturers to prevent "short cycling" when doors are opened too long or excessively. A temperature lag between the feeler tube and the evaporator plate is thermal barrier.

Also, this arrangement prevents chemical reaction between the feeler tube and the aluminum plate.

SERVICE DIAGNOSIS - - - SELF-DEFROSTING REFRIGERATORS

Refrigeration has become an indispensable part of the average person's daily life. Any interruption in the normal service rendered by the refrigerator is keenly felt by the user. Consequently, the personal contact which the serviceman has with the customer, for whom the service is rendered, is of primary importance along with the actual service performed.

To properly (and successfully) service refrigeration equipment, these are several fundamental factors to guide the serviceman.

A thorough understanding of the theory of refrigeration.

A good working knowledge of the purpose, design and operation of the various mechanical parts of the refrigerator.

The ability to diagnose and correct any trouble that may develop.

This discussion deals primary with diagnosing trouble in self-defrosting refrigerators, through a step-by step procedure. When the analysis of the trouble has been completed and properly determined, apply that correction in a straight-forward manner.

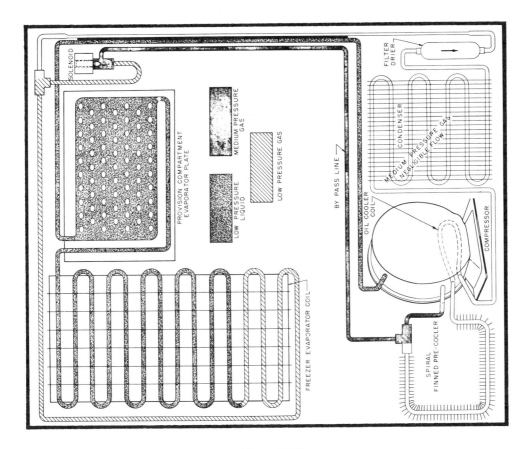

Figure 130

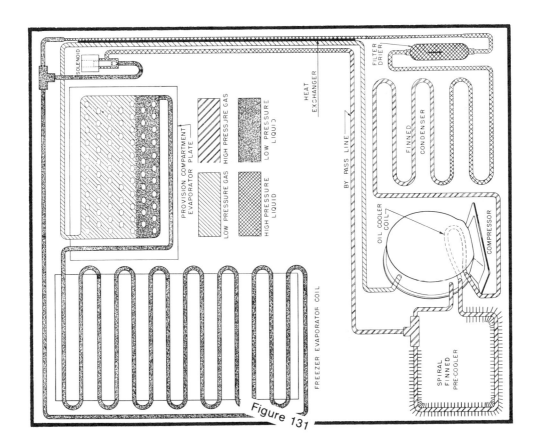

Figure 131

Always allow the customer to explain the problem. Many times the trouble can be diagnosed more quickly, based on the customer's explanation. Most of all, do not jump to conclusions, until you have heard the full story and have evaluated the information obtained from the customer.

Before starting a test procedure, connect the refrigerator service cord to the power source, through a wattmeter, combined with a voltmeter. Then, make a visual inspection and operational check of the refrigerator to determine the following:

Is the refrigerator level?

Is the refrigerator (static condenser model) located for proper dissipation of heat from the condenser? Recommended spacing 3½'' from rear wall, and 4'' clearance above cabinet.

Are the door gaskets sealing on pilaster area.

Do the provision and freezer compartment doors actuate the fan and/or light switch?

Are the fan blades properly located on the motor shaft?

Is the thermostat thermal element properly positioned in the thermal well? Thermal element must not touch the humidiplate.

What position is the thermostat dial set on?

Observe the frost pattern on humidiplate.

Check the condenser for lint and dust. Is the shroud in place? (Forced convection models)

Feel the condenser. with compressor operation, the condenser should be hot, with a gradual reduction in temperature from the top to the bottom of the condenser.

Inscribe the bracket opposite the dial or slotted shaft of the defrost control to determine if the control advances. *(NOTE: Some defrost controls advance only when the compressor operates).*

Is wire food storage rack in bottom of freezer compartment reversed? A reversed rack may restrict air circulation.

Are defrost control thermal element retainer clips in place across the bottom of the freezer coil if used?

Is the drain trough heater inserted down into the freezer drain tube and clear of the defrost thermal element? (on refrigerators so equipped)

After this phase of diagnosis is completed, a through operational check should be made of the refrigeration system and any components not previously checked.

Thermostat Cut-Out and Cut-In temperatures.

Freezer and provision compartment air temperatures.

Line Voltage.

Compressor Wattage.

Compressor Efficiency.

Refrigerant Charge.

Restrictions.

Defrost Solenoid Valve Operation.

Defrost Control Termination Temperature.

THERMOSTAT CUT-OUT AND CUT-IN TEMPERATURES

To accurately check the cut-out and cut-in temperatures of the thermostat, use a refrigeration tester, or a recording meter. Attach a refrigeration tester or recording meter bulb firmly to the thermal well on the humidiplate. (On models with evaporator covers, replace the cover before starting test.)

When using a refrigeration tester equipped with several bulbs, place No. 2 bulb in Food Compartment air (center), No. 3 bulb in Freezer Compartment (center), No. 4 bulb on defrost control thermal element. Refer to item Defrost Control Termination, for proper location of bulb.

Allow the system to operate through a complete cut-out and cut-in cycle, with the thermostat set on the middle position.

It is essential that the temperature of the thermostat thermal element, not be reduced more than one degree per minute, through the final five degrees prior to cut-out, for accurate reading or recording.

Erratic operation of the thermostat will affect both the Freezer Compartment and Food Compartment air temperatures.

MOTOR OVERLOAD AND DEFROST CYCLE.

The motor overload protector may trip during the defrost period: When the defrost cycle is initated, the solenoid valve opens to allow the hot compressed refrigerant vapor to by-pass the condenser and capillary tube. In case the compressor has been running and is quite warm at the start of the defrost cycle, the discharge pressure will drop and the suction pressure will increase rapidly. The pressures, however, do not become balanced due to frictional resistance in the lines, solenoid valve, in the compressor muffler, etc. The result is high suction pressure and a pressure differential which results in a very high motor load. The refrigerant returning to the compressor cools the motor quite rapidly, however, the motor overload senses only the high motor load (current), and does not immediately sense that the motor is cooling. As a result, the overload trips.

The overload resets very soon and the compressor again runs. This cut-out on the overload may be repeated several times. Cycling of the compressor on the overload does not stop the defrosting of evaporators, as sufficient hot refrigerant vapor is pumped into the evaporator to continue the defrosting, even while the compressor is stopped. Tripping of the overload does no harm, and merely indicates that the overload device is doing its job of protecting the motor. A temperature lag between the motor windings and the overload, results in tripping of the overload when certain conditions are present. These conditions are high room temperature or machine compartment temperature, a hot compressor which has been

running for some time, or an evaporator which has little or no frost accumulation. When there is little or no frost at the beginning of a defrost cycle, the suction pressure (or load) rises rapidly, faster than the overload can feel the temperature of the cooler motor windings.

When the defrost control thermal element temperature reaches approximately 43° F, this temperature sensitive element opens the electrical contacts in the defrost control to close the solenoid valve.

The compressor continues to operate, to lower the freezer and provision compartment evaporator temperatures to normal operating range. It then cycles to maintain normal storage temperatures in the refrigerator.

The electric solenoid valve in the by-pass line normally draws 6 watts at rated voltage. A stuck closed solenoid valve will draw approximatley 10 watts.

This specification must be checked for each model.

CHECKING HIGH TEMPERATURE COMPLIANTS WITH A WATT METER

It is possible that a compressor that is drawing too much wattage may run and refrigerate but it is not refrigerating properly. It may not reach temperatures satisfactory to shut off the control of the unit. Extreme running times will be the result. The use of a watt meter where it can·be plugged into the wall receptacle and the refrigerator plugged into the watt meter will insure a proper check being made. If the wattage reading is excessive, compared to the manual specifications, it is a very good indication that the compressor should be changed.

LINE VOLTAGE

It is essential to know the line voltage at the appliance.

A voltage reading should be taken at the instant the compressor starts, and also while the compressor starts, and also while the compressor is running. Line voltage fluctuation should not exceed 10%, plus or minus, from nominal rating. Low voltage will cause overheating of the compressor motor windings, resulting in the compressor cycling on thermal overload, or the compressor may fail to start.

Inadequate line wire size, and overloaded lines, are the most common reasons for low voltage at the appliance.

WATTAGE

Wattage is a true measure of power. It is the measure of the rate at which electrical energy is consumed. Therefore, wattage readings are useful in determining compressor efficiency, proper refrigerant charge, if a restriction exists, or if there is a malfunction of an electrical component.

Amperes, measured with an Amprobe, multiplied by the voltage, is not a true measurement of power in an alternating current (AC) circuit. It gives only "volt-amperes" or "apparent power".

This value must be multiplied by the Power Factor (phase angle), to obtain the true or actual (AC) power factor. The actual power is indicated by a wattmeter.

Expressed:

$$Watts = Volts \times Amperes \times Power\ Factor$$

or

$$Power\ Factor = \frac{Watts}{Volts \times Amperse} = \frac{Actual\ Power}{Apparent\ Power}$$

Thus, Power Factor may be expressed as the ratio of the actual watts to the apparent watts. The apparent watts is the product of the amperes and volts as indicated by an ammeter and a volt-meter. Power Factor varies from zero (for a pure reactance) to 1.00 (or 100%) (for a pure resistance). On resistance heaters, such as the drier coil and the drain heater, the actual watts are equal to the amperes multiplied by the volts, thus, the Power Factor is 100% or 1.00. With electric motors, because of their magnetic reaction, the actual watts are not equal to amperes multiplied by volts, and the Power Factor is less than 1.00.

It is for this reason a wattmeter should be used.

COMPRESSOR EFFICIENCY

An inefficient compressor causes excessive or continuous compressor operation, depending on the ambient temperature and service load. Recovery of cabinet temperature will be slow, if cycling does occur. Wattage will be below normal. Suction pressure will be slightly above normal, and will remain constant with continuous operation. Frost accumulation will be soft. Suction and discharge pressures will balance very quick on the off-cycle. Condenser temperature will be near normal.

If the compressor has service valves, turn the suction service valve "in" (CW) and pump compressor down to a "vacuum". Stop, compressor and observe suction pressure gauge reading. A rise in pressure indicates discharge valve reeds are leaking.

REFRIGERANT SHORTAGE

A loss of refrigerant results in excessive or continuous compressor operation; above normal Food Compartment temperature; a partically frosted humidiplate (depending on amount of refrigerant lost); below normal Freezer Compartment temperature; low suction pressure (vacuum); and low wattage. The condenser will be "warm to cool", again depending on the amount of refrigerant lost.

When refrigerant is added, the frost pattern will improve; the suction and discharge pressure will rise; the condenser become hot; and the wattage will increase.

The refrigerator should be turned off and throughly leak tested.

RESTRICTIONS

Restrictions are classified as follows: Permanent; total or partial, as a result of foreign matter; oil or moisture in the capillary tube, strainer or drier.

A permanent total restriction completely stops the flow of refrigerant through the system. The result is, continuous compressor operation, low wattage, low suction pressure (vacuum), and a cool condenser which indicates liquid refrigerant is trapped in the condenser.

A partial restriction results in a partially frosted humidiplate, excessive or continuous compressor operation, above normal Food Compartment temperature, below normal Freezer Compartment temperature, low wattage and low suction pressure (vacuum), depending on the amount of restriction. The low (or outlet) half or one-third of the condenser will be cool, indicating liquid refrigerant trapped in the condenser.

To make sure that the trouble is a partial restriction and not a shortage of refrigerant, cover the condenser. Allow the compressor to operate to increase the discharge pressure and temperature (note increase in wattage). The increase in discharge pressure will force refrigerant through the restricted area, and frosting of the humidiplate will occur. The frost pattern will not improve with a shortage of refrigerant.

A total or partial "moisture" restriction always occurs at the outlet of the capillary tube. If moisture is suspected, turn the refrigerator "off" and allow all system temperatures to rise above 32° F, or manually initiate a defrost cycle. If moisture is present, the restriction will be released.

DEFROST SOLENOID VALVE

The defrost solenoid valve must be placed in a true vertical position to prevent friction between the armature and armature tube, when valve opens and closes.

An improperly seated defrost valve needle permits high pressure refrigerant vapor to leak into the Freezer Compartment evaporator coil, causing a rise in temperature and reduced coil efficiency. This results in additional compressor running

time. A minute leakage is not detectable by a rise in temperature of the by-pass line, in normal ambients, since the rate of leakage is less than the rate at which the vapor condenses in and by-pass line, and the temperature remains constant. As a result, only liquid refrigerant enters the Freezer Compartment evaporator coil. Such leakage will generally not affect freezer temperatures.

Should the leakage rate exceed the rate at which the refrigerant vapor condenses in the by-pass line, high pressure refrigerant vapor would then enter the Freezer Compartment evaporator coil, resulting in a four degree, or higher, rise in average Freezer Compartment temperature. The temperature rise will be determined by the quantity of vapor entering the coil through the leaking solenoid valve. The by-pass line temperature will rise (detectable by touch) and the suction line will be cold (below room temperature at cabinet exit). Wattage would be above normal, with increases or continuous compressor running time. Should cycling occur, system pressures would equalize rapidly on off cycle.

DEFROST CONTROL TERMINATION TEMPERATURE

Before attempting a termination temperature check, it is essential to know, that the guide tube heater, and the drain sump or drain trough heaters are operating properly, in their respective models, since the defrost control termination temperature is influenced by their operation.

To obtain the most accurate termination temperature reading, there are specific locations at at which the readings must be taken. Always refer to manufacturer s specifications!

In some models with the thermal element inserted in the fins of the evaporator coil, or in the thermal well on the rear edge of the evaporator, attach a refrigeration tester bulb to the defrost control thermal element, between the drain trough and the freezer evaporator.

In others, with the defrost control thermal element in the drain trough only, secure the refrigeration tester bulb on the defrost control thermal element in the drain sump or in the drain trough.

In still others insert the refrigeration tester bulb down into the freezer compartment rubber drain tube. The lead end of the thermal bulb should be level with the top of the drain tube, and in contact with the thermal element.

After the refrigeration tester bulb has been positioned, allow the system to operate for a short time, to allow the thermal couple bulb to cool. Initiate a defrost cycle manually. Record the defrosting time, wattage and defrost control termination temperature.

The temperature of the inlet refrigerant can be taken with a Refrigeration Tester bulb attached to the Freezer Compartment evaporator coil inlet line. A refrigerator which is designed for an average freezer air temperature of $0^{\circ}F$, will have a refrigerant inlet temperature of 8° to 14° below the design average air temperature, at cut-out.

AIR FLOW PATTERN

Air flow patterns in modern refrigerators vary widely and in each case is established by the cabinet design and the method used. Some are forced air flow while others depend on simple convection principle. One model may just have

one fan, others two fans. Our sample flow system has one fan to simplify diagnosis of the system.

The one fan circulates air to both the food area and the freezer compartment. Here, the design is chosen as a means of balancing the air and to mix the air from the freezer and food area as it returns to the evaporator.

Figure 132 shows the circulation paths to and from both compartments.

Starting at the fan, the air is pulled up through the evaporator coils into the fan blade where it is blown out into the freezer area.

Settling back down to the air flow baffle in the bottom of the freezer compartment some of this air flow is directed back to the evaporator to repeat the cycle, *Figure 133*

However, the diffuser plate directly opposite the fan will force some of the air flow back into the air duct leading to the food compartment. Then it eventually returns to the evaporator after pas-

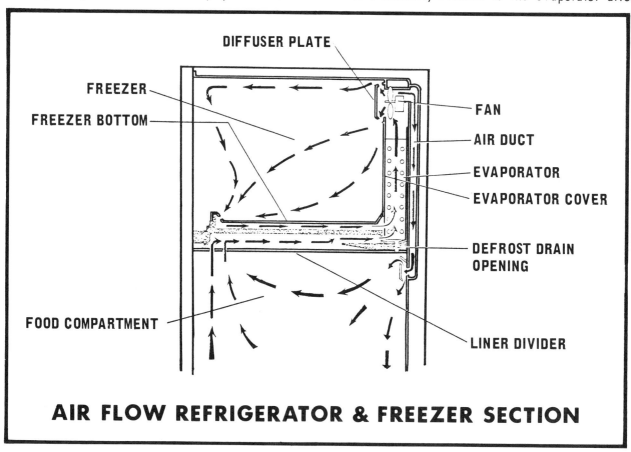

AIR FLOW REFRIGERATOR & FREEZER SECTION

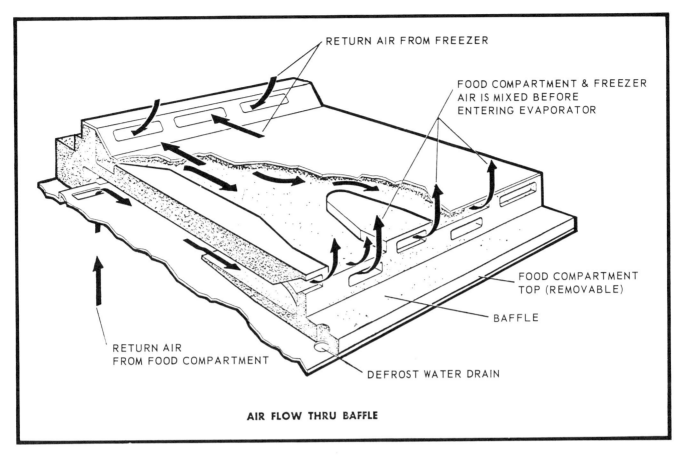

RETURN AIR FROM FREEZER

FOOD COMPARTMENT & FREEZER
AIR IS MIXED BEFORE
ENTERING EVAPORATOR

FOOD COMPARTMENT
TOP (REMOVABLE)

BAFFLE

RETURN AIR
FROM FOOD COMPARTMENT

DEFROST WATER DRAIN

AIR FLOW THRU BAFFLE

Figure 133

sing through the bottom section of the air flow baffle shown in *Figure 133* Here it is mixed by the fan's action and the heat removed.

The fan runs only when the unit control is on, the freezer door closed and the defrost heater is off.

Air flow may be redirected to a "Meat Keeper" where the air flow is split as shown in *Figure 134*

Also there is yet another air flow pattern used where the action of the fan, with the blade towards the back, causes just the reverse flow. It pulls the air up through the evaporator, through the air duct. Passing right and left, it is blown back into the evaporator through the freezer air diffuser. Some of the air posses downward to the refrigerator air diffuser outlet. Again part of the air flow reaches the bottom of the food compartment by special duct and covers the meat keeper, *Figure 135*

With these two examples of an air flow it is possible to appreciate why the serviceman must always check to see that the customer doesn't cover up or obstruct these ducts and openings to

interfere with the air flow. Checking for this flow path in any refrigerator is of first importance because it can give adnormal diagnosis results and waste time running down the trouble.

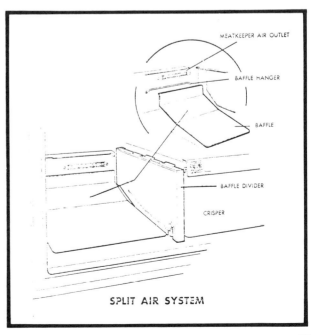

MEATKEEPER AIR OUTLET

BAFFLE HANGER

BAFFLE

BAFFLE DIVIDER

CRISPER

SPLIT AIR SYSTEM

Figure 134

117

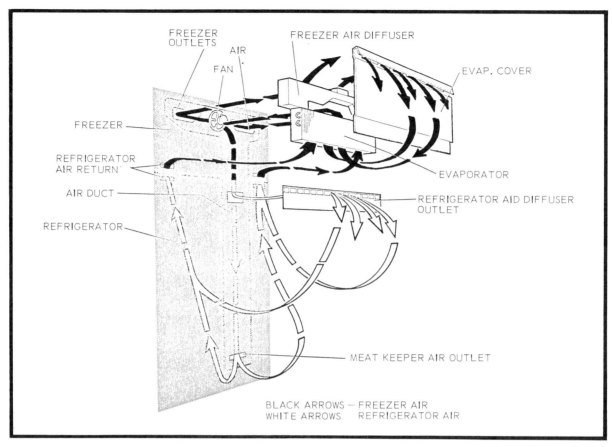

FREEZER OUTLETS
AIR
FAN
FREEZER AIR DIFFUSER
EVAP. COVER
FREEZER
REFRIGERATOR AIR RETURN
AIR DUCT
REFRIGERATOR
EVAPORATOR
REFRIGERATOR AID DIFFUSER OUTLET
MEAT KEEPER AIR OUTLET

BLACK ARROWS — FREEZER AIR
WHITE ARROWS — REFRIGERATOR AIR

Figure 135

EPOXI REPAIRS.

The use of a new modern material called "*EPOXI*" is of great aid to the refrigeration mechanic. It comes in white, aluminum, black, light blue and light green, as well as in clear and off-white. An approved kit of this patch substance (such as "*HYSOL*" Epoxi Patch")will also repair chipped enamel, porcelain, plastisol liners, etc. as well as make repairs on pipes, tubing and fittings.

If the leak in the evaporator has been caused by customer carelessness or by other means, there is a quick way to make a repair. First, attach a drum of R–12 refrigerant to the low side of the system, using a piercing valve. There should be a gauge in series with the drum and a pressure of

10 to 15 pounds kept on the system during the repair. This will avoid introducing foreign materials or more moisture into the line.

Next, sand, carefully, the area around the hole and thoroughly clean the area with acetone. Now, the pressure can be removed from the system.

Mix the epoxi as directed on the tube. Equal lengths of resin and hardener should be mixed for about two or three minutes and immediately applied. An amount the size of a dime will usually suffice. To cure the patch, apply the heat from a 250 watt lamp and this will also speed up the process to somewhere around two hours time for a finished job. The job may be touched up with a spray paint to match the surrounding area.

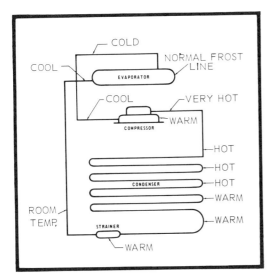

NORMAL RUNNING UNIT

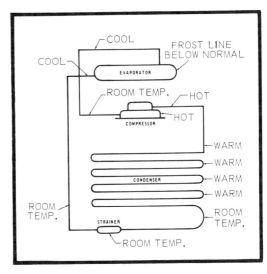

REFRIGERANT SHORTAGE

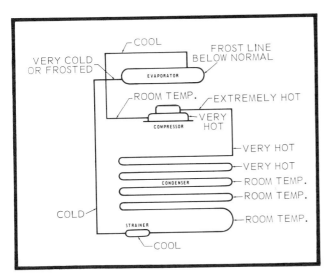

PARTIALLY RESTRICTED STRAINER

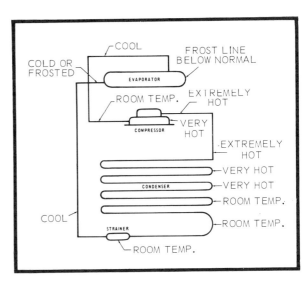

PARTIALLY RESTRICTED CAPILLARY

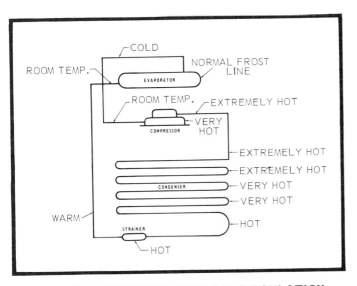

ACROSS CONDENSER POOR AIR CIRCULATION

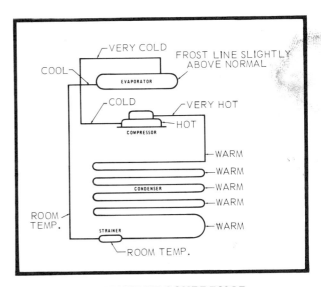

INEFFICIENT COMPRESSOR

Figure 136

TOUCH METHOD OF DIAGNOSIS (See Figure 136)

The "touch method" of diagnosis has been used by experienced servicemen for many years and has been proven very valuable as a fast preliminary check. The above diagrams and following explanations refer to units that are running in a normal ambient of 75° F. and are factory originals-that is, the system was working properly prior to the malfunction.

Due to the many variables involved, exact temperatures cannot be given for the different test points shown. Your own experience will teach you to judge these temperatures.

Compressor
Body temperature tells you the running time and load. Warm or very warm is normal; very hot would indicate excessive running time or load.

Condensor
Temperatures should gradually decrease from the compressor discharge to the strainer. A sudden drop to room temperature indicates a liquid build-up, usually as a result of a restricted capillary tube or strainer.

Strainer
Temperatures are normally about 5 to 10 degrees above room temperature. A cool or cold strainer indicates a restriction. An excessive load on the evaporator or lack of air circulation at the condensor would make the strainer very warm or hot.

Capillary
Tube temperatures should be at room temperature near the point of heat exchange to the suction line, and cool near the evaporator. A sudden drop in temperature along the capillary tube indicates the point of restriction.

STRAINERS
All sealed refrigeration systems have a strainer in the liquid line between the condenser and the capillary tube, *Figure 137* On some models it may be a combination strainer and drier referred to as a filter-drier.

The strainer has one, or a series of, fine mesh screens placed inside a copper tube in such a manner that the liquid refrigerant must pass through them before entering the capillary tube.

This prevents any foreign particles in the system from entering the capillary tube and causing a restriction.

In most cases the manufacturer simply fastens a sleeve shaped piece of screen in the end of the liquid line where it fastens to the capillary tube. In some cases approximately two inches of the end of the tube may be enlarged slightly to accommodate a larger screen.

Any time a capillary tube or heat exchanger is replaced, the strainer should also always be replaced. If a capillary tube has a restriction, it can readily be assumed that the strainer has failed and allowed foreign material to pass. A restricted strainer can be easily detected by feeling it while the unit is running. If the strainer feels cool to the touch, it is restricted or partially restricted.

Figure 138 shows two types of strainers. One is a strainer by itself with one end spun down to fit a capillary tube. These ends are available in different sizes according to the application. The second strainer shown in *Figure 138* is a combination strainer and capillary tube.

When using a combination strainer and capillary tube or, whenever replacing the capillary tube by itself, the capillary tube should be soft-soldered to the suction line. Failure to do this results in the loss of the heat exchange principle and thus lowers the efficiency of the refrigeration system. It will be found that without the heat exchanger, the system appears to function properly but, under peak load conditions, the efficiency will be sharply reduced.

Replacing a strainer involves the same procedure used for any other component. Attach a piercing valve on the process tube of the compressor and purge all the refrigerant from the system. Carefully clean the tubing behind the strainer and cut the tubing with a tube cutter. If the strainer has tubing at the outlet end, follow the same procedure for this tube. If the capillary tube is soldered into the strainer outlet, clean the capillary two or three inches from the strainer, score the tube with the edge of a file and carefully break it off.

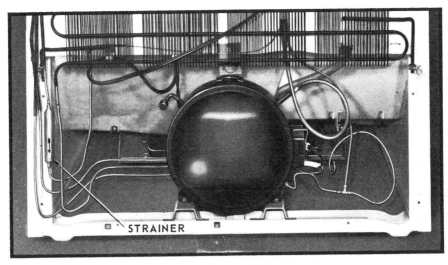

Figure 137

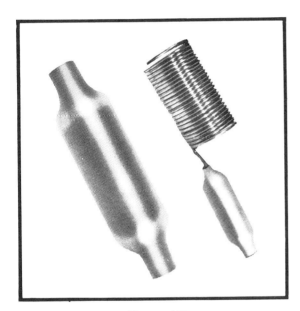

Figure 138

Insert the liquid line and capillary tube into the new strainer and silver solder both joints. Evacuate and recharge the system following the procedure outlined in this manual.

COMPRESSOR TERMINALS

Sealed units, with the compressor and motor inside a steel shell, must have some means of conducting electricity inside the shell in order to operate the motor. Since the starting device or relay is outside the shell, there are three wires that must be brought through the shell. The three wires from the motor are connected to terminals which are brought through, and insulated from, the shell itself. These three terminals are then the common, start and run terminals which the external wires from the relay are connected to.

Most late style compressors have the three terminals molded into a piece of glass or "Fusite". This glass is mounted into the compressor shell and acts as the insulation between the terminals and the steel shell. If this type of terminal develops a leak or if one of the terminals becomes damaged, the compressor must be sent to a rebuilding shop for repair.

Earlier style compressors have "built up" terminals. These are threaded terminals with two or more nuts and a series of fiber and steel washers to make up the complete terminal. Quite often this style terminal may develop a leak where it goes through the shell. This type leak can be quickly and easily repaired in the field, provided certain precautions are taken.

When a terminal seal leak is encountered, remove the lead wires connected to the terminals. Next, remove the remaining nuts and washers down to the compressor shell. In all cases, even if only one terminal is leaking, install new seals on all three terminals.

When removing the bottom nut, care must be taken that the terminal does not turn. If this happens, the wire inside the shell can be broken off. The compressor would then have to be taken to a rebuilding shop. Always use an open end or small box wrench and watch the terminal while slowly turning the nut. If the terminal starts to turn, stop turning the nut immediately and try the corrective measures outlined below.

If the nut is not too tight, many times a light oil or penetrant on the terminal threads will allow the nut to turn more freely. If oil does not help, take two of the terminal nuts previously removed when removing the lead wires and screw them onto the end of the terminal. Tighten the two nuts against each other securely (turn one clockwise and the other counterclockwise). You can now hold these two nuts with a wrench to prevent the terminal from turning while removing the nut. This will not damage the threads. Never use a pair of pliers directly on the threads to hold the terminal.

In extreme cases where you cannot remove a nut from the terminal, there is a tool available called a "nut buster". The tool is placed over the nut and turning a threaded shaft drives a chisel point through the edge of the nut. After the nut is split, it can be easily removed.

After all the nuts and washers are removed, carefully clean the compressor shell of all paint, dirt, etc., from around the terminals. There are several different kits on the market to repair the terminals and all contain enough material to repair all three terminals. While there are several different kits, there are only two basic styles.

One style, *Figure 139* is constructed in one piece and screws down on the old terminal. A gasket on the bottom of the seal mates against the compressor shell forming an effective seal. Extreme care must be used when installing this style in order to make sure that the terminal does not turn. There is no way to hold the terminal, so if the seal does not turn freely all the way down, do not use this style!

The second style terminal seal is shown in *Figure 140* These style consists of a neoprene seal that seals against the compressor shell with a series of washers and nuts to complete the repair. With this style, the terminal can be held from turning while installing the seal.

When making a terminal seal repair, do not use any type of glue or cement to aid in sealing. It is not necessary and will cause a future leak. After the repair is made, evacuate and recharge the system.

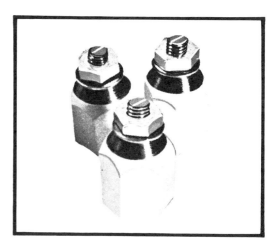

Figure 139

Figure 140

REPLACEMENT MAGNETIC LATCHES

Early model refrigerators and freezers frequently have the door latch fail. Many times it is difficult or even impossible to obtain a replacement. Many of these latches are also quite expensive.

There is available, a universal replacement latch that is magnetic in operation. This latch is available at most parts distributors and in many cases, it will cost much less than the original replacement.

The latch, shown installed in *Figure 141* also has the added advantage of being child-safe. This means that the door can easily be opened from the inside merely by pushing on it, should a child become trapped inside an empty or abandoned refrigerator.

When installing the latch (trade name Magnalatch), the old strike should be removed and the handle fastened so that it remains firm when pulling on it. If the handle is removed or missing, there is a universal handle also available. This handle is trade named Adaptagrip and is also shown in *Figure 141.* This handle will not cock the old style latches, so it cannot be used unless the latch is the self cocking type magnetic, or the refrigerator has a magnetic gasket.

Both the latch and handle come with simple instructions for easy mounting on any refrigerator or freezer. Their ease of installation and low cost makes them especially suited to the rebuilding trade.

REFRIGERANTS

The chart shown in *Figure 142* lists the characteristics of some of the more popular refrigerants encountered in refrigeration service work. Listed are the ARSE number, the chemical name, chemical symbol and the boiling point of the refrigerants.

A refrigeration system engineered to operate with one type of refrigerant will not operate properly

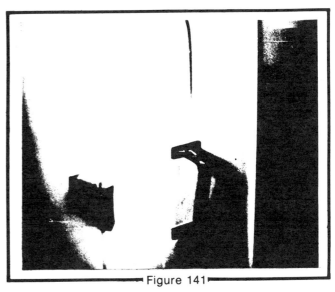
Figure 141

with any other type. Always be certain before adding a refrigerant to a system that the proper one is used. The manufacturer lists the correct refrigerant to be used on the serial number plate of the unit or cabinet. This may be listed according to the ARSE number, the chemical name or the chemical symbol. It is important therefore, that the information on the chart be noted so that you will know the correct refrigerant to use, no matter how it is listed on the serial plate.

R–11 Trichloromonofluoromethane CCl_3F B.P. 74.7° F.	Used as a refrigerant in industrial and commerical air-conditioning systems; also, in industrial process water and brine cooling to $-40°$ F. ($-40°$ C.). Also used as a solvent for cleaning up a refrigeration unit which has experienced a burnout in the hermetic motor.	**R–113** Trichlorotrifluoroethane CCl_2F-$CClF_2$ B.P. 117.7° F.	Used as a refrigerant in industrial and commerical air-conditioning systems; also, in industrial process water and brine cooling to $0°$ F. ($-17.8°$ C.).
R–12 Dichlorodifluoromethane CCl_2F_2 B.P. $-21.6°$ F.	Used as a refrigerant in both direct and indirect industrial, commerical, household and automotive air-conditioning systems; also, in household refrigerators, ice cream cabinets, frozen food cabinets, food locker plants, water coolers, etc.	**R–114** Dichlorotetrafluoroethane $CClF_2$_$CClF_2$ B.P 38.4° F.	Used as a refrigerant in fractional horse-power household units and drinking water coolers employing rotary compressors; also in industrial and commerical air-conditioning systems and in industrial process water and brine cooling to $-70°$ F. ($-56.7°$ C.).
R–13 Monochlorotrifluoromethane $CClF_3$ B.P. $-114.6°$ F.	Used as a refrigerant in both direct and indirect industrial very low-temperature cascade systems ranging in size from fractional to 100 horsepower.	**R–500** Azeotrope CCl_2F_2 - CH_3CHF_2	An azeotrope of R12 refrigerant which has slightly higher vapor pressures and provides higher capacities from the same compressor displacement.
R–22 Monochlorodifluoromethane $CHClF_2$ B.P. $-41.4°$ F.	Used as a refrigerant in room air-conditioners, central residential air-conditioners and heat pumps employing reciprocating or rotary compressors; also in industrial and commerical low-temperature refrigerating systems employing reciprocating compressors.	**R–502** Azeotrope $CHClF_2$ - $CClF_2\,CF_3$	An azeotrope of R22 refrigerant which is especially suited to low evaporator temperatures. Can be used equally as well with high temperatures. Provides capacity gains.

DOOR GASKETS

The necessity of a tight door seal cannot be over emphasized. Before any adjustments are made on the cabinet, first and foremost, level the cabinet. A cabinet that is not level can have poor sealing of the door gasket. This is particularly true with magnetic door gaskets.

With the cabinet level, the most practical method of checking the door seal, is with a light. This is more exacting than using paper currency, business cards, etc. Make up an extension cord using a 100 to 150 watt lamp. The cord should be made from small thin fixture wire to permit easy closing of the door on the wire. Many servicemen install a piece of flat TV lead-in wire in the cord for this purpose. Locate the lamp on the center shelf, as near the front as possible. Use care to avoid contact with plastic parts. Close the door and darken the room as much as possible. The slightest leak at the door gasket can be detected by visible rays of light. The light rays can be seen through an opening that would exert a slight pressure on a piece of paper .001'' thick.

Door Adjustment and Alignment

Poor door seal is, in most cases, caused by hinge bind, improper strike adjustment or the door not in alignment with the front of the cabinet. In the event a door does not have a proper fit, first check it for hinge bind. Hinge bind may occasionally be found, resulting in excessive compression of the door gasket at the hinge side, which prevents the gasket from sealing properly against the pilaster at the latch side. To relieve hinge bind, or to improve gasket seal at the hinge side, the door must be adjusted away from, or closer to, the cabinet as required. On some refrigerators, this is accomplished by adding or removing shims under the hinge. On other models, the hinge screw plates inside the cabinet are movable to allow for adjustment. Loosen the screws in the hinge butt plate and move the door in or out to effect the proper door seal.

In the event the door is not in alignment with the front of the cabinet, it may be realigned as follows:

For a slight realignment, it is not necessary to remove the inner panel. Loosen all the gasket retaining screws one-half to one turn except two at each corner. Slightly loosen the corner screws until the door can be moved by exerting an outward pull on the tight corner.

Try the door for a proper seal, and continue to move the tight corner out and the high corner in until the door gasket is square with the cabinet pilaster. Tighten firmly all the retaining screws and recheck the door seal.

After the door has been properly aligned, it may be necessary to adjust the door strike to effect a proper seal. Loosen the strike adjusting screws and move the strike in or out as required.

If, after the above adjustments are made, light still passes through the door gasket at certain points, fill in behind the gasket with strips of tape or similar material. Place the tape directly under the balloon portion of the gasket and on top of the inner panel.

Replacing Door Gasket

The refrigerator door gasket is fastened to the door flange by self-tapping screws placed through holes in the plastic inner door panel. To replace the gasket, remove the inner door panel. Press the gasket out with the thumb to expose the retaining screws and remove the screws, *Figure 142*. Remove the inner panel and gasket as an assembly and remove the old gasket from the inner panel. Install the new gasket on the inner panel and pull and stretch the gasket until it fits the panel smoothly.

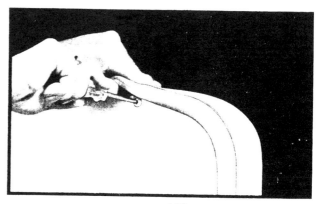

Figure 142

In some cases, the replacement gasket may be a straight length instead of a welded frame. In these cases, start the gasket at the middle of the bottom edge of the inner panel. Using masking tape, tape the gasket at intervals around the panel as you install the gasket. Loop the tape from the back of the panel, around the gasket and to the front side of the inner panel. If the gasket does not have notches for the corners, you will have to carefully cut these notches as you install the gasket.

After the gasket has been properly installed on the inner panel, install the inner panel and gasket to the outer door panel. Install only two retaining screws at each corner and tighten only slightly. Align the door as described under "Door Adjustment and Alignment". Install the remaining retaining screws and check for a proper seal. Do not tighten the retaining screws too tightly, as this can crack or warp the inner panel. If the gasket has been taped to the inner panel, cut the exposed tape off and discard.

DRIER

ALWAYS INSTALL A NEW DRIER WHENEVER THE SEALED SYSTEM IS ENTERED. A factory compressor often includes a new drier.

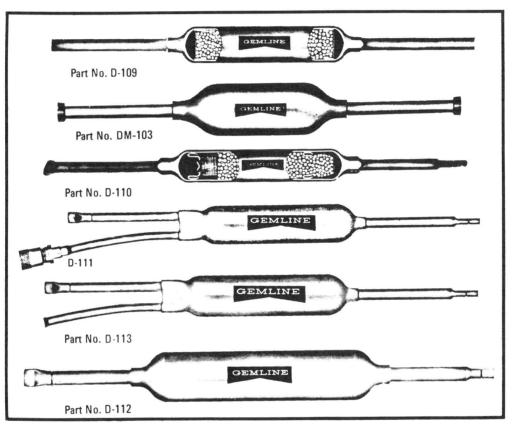

Part No. D-109

Part No. DM-103

Part No. D-110

D-111

Part No. D-113

Part No. D-112

Figure 143   *Courtesy of Gem Products, Inc.*

COPPER EXTENDED END DRYERS					DROPS OF WATER REMOVED				RECOMMENDED TONNAGE			
					R-12		R-22		R-12		R-22	
Part No.	Inlet	Outlet	Dia.	Length	75^0	125^0	75^0	125^0	75^0	125^0	75^0	125^0
D-109	1/4	1/4	3/4	9	36	36	39	29	3/4	3/4	1/2	1/3
D-110	1/4	cap. & 1/4	3/4	9	36	36	39	29	3/4	3/4	1/2	1/3
D-111 & D-113	1/4	cap. & 1/4	1	9-7/8	76	72	62	46	1-3/4	1-1/2	1	3/4
D-112	5/16	cap. & 5/16	1-3/16	10½	108	108	93.3	69.6	2.5	2.5	2	1.5
DM-103	1/4	1/4	1	8	36	36	31	23	3/4	3/4	1/2	1/3

Do not install the new drier until all of the other repairs have been made. If you are using a drier with a Schrader access fitting remove the valve core before brazing. The drier should be positioned with the capillary tube side lower than the inlet to the drier. This will assure a full supply of liquid refrigerant to the capillary during operation.

See text REPLACING OR INSTALLING A DRIER. The original drier should be removed from the system, use your cutting tool or a hack saw blade. Do not attempt to use the torch to remove the drier. Excess heat will drive the moisture trapped in the old drier back into the system.

There are many service dryers to select from, with each model offering optional features and capacities, *Figure 143*

Those dryers illustrated incorporate the popular all molecular sieve bead desiccant, which offers through its high absorption qualities, the ability to hold several times the water of other absorbants. Before attempting to install a drier, read the text EVACUATING THE SYSTEM. The drier is installed in the last pass of the condenser where the capillary tube enters on the high side. If you have a static type condenser in the back of your refrigerator, it may be necessary to break the tubing away from the main frame of the condenser to install the drier.

1. After the system has been evacuated and the pressure balanced with refrigerant, cut the capillary tube at the old drier or strainer with a side cutter or dykes.

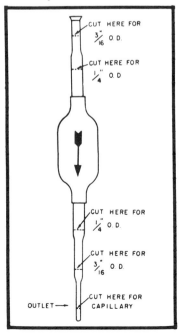

2. The last pass of the condenser may have a very small strainer in the tube ahead of where the capillary tube was brazed. Be aware of this and remove this screen, even if part of the condenser must be eliminated. A regular strainer or drier is easily recognized and should be removed.

3. Clean and polish the tubing where the brazing will take place on the condenser, also clean and polish the end of the drier.

4. Fit the drier to the condenser, use a swaging tool to expand the tubing on the drier if necessary.

5. Fit the capillary tubing into the drier.

6. When both ends are made to fit, you are ready for brazing.

7. Remove the capillary tube from the drier.

8. Using a good flux, coat the joint that is to be made.

9. Now slip the drier to the condenser tube just a bit.

10. Light torch and braze, when metal rod starts to flow, push the drier into the condenser tube as far as it will go and complete the brazing.

11. With a file cut the end of the capillary tube at an angle, do not file straight through. Make a diagonal cut on both sides *Figure 144,* then with a pair of pliers, break the cut off. The result should be a smooth elongated hole.

Figure 144 — *Capillary Tube Preparation*

12. Place some flux about an inch away from the end of the tube. With your torch, allow some brazing rod to flow around the capillary tube at the point the flux was spread.

13. Install capillary tube in drier, heat and melt the metal as you push the capillary tube in, complete the brazing.

CAUTION
All valves should be open to the atmosphere when brazing. If pressure builds up a leak will result

REFRIGERANTS AND REFRIGERANT OIL

Mechanical refrigeration equipment in use today is dependent upon a fluid called REFRIGERANT to perform its task in the removal of heat. The refrigerant pumped through the system is recycled constantly. It is never consumed. Replacement is only necessary in the event of a leak, or repair to the system or the equipment.

The refrigerant becomes saturated with heat removed from the confined area that is to be cooled or refrigerated and transferred to the condenser where the heat is released to the ambient air. A refrigerants ability to vaporize at low temperatures makes it a desirable refrigerant. The refrigerant in an operating system is constantly changing its state from liquid to vapor then from vapor back to liquid but never changing its chemical structure. The ideal refrigerant must not react chemically with oil, nor decompose under normal working temperatures and pressures. Refrigerants must not attack metals, and they should be safe to use in a system and in handling.

In the following text you will learn what is required of a desirable refrigerant, and become acquainted with the fluorocarbon families of refrigerants and their adaptibility, and of the special purpose refrigerants such as ammonia and R 502. Selecting the proper refrigerant for special purposes will also be explained.

EARLY DEVELOPMENT OF REFRIGERANTS

Fluids used in air conditioning and refrigeration are called refrigerants. These refrigerants have identifying numbers and the first letter R in the word refrigerants is placed before the number for identification of the refrigerant. R 12, R 22, and R 502 are a few of the designated numbers. The word "Freon" is still used by refrigeration service people to describe refrigerants. Freon is a registered tradename used by one of the many manufacturers of refrigerants.

Other tradenames by various manufacturers, Genetron, Isotron, and Ucon are a few of the more popular brand names. Early refrigeration equipment employed sulfer dioxide, methyl chloride or ammonia as a refrigerant. These are still good refrigerants but unsafe, both in the system and in handling. Because the refrigeration industry was in need of a safe refrigerant with desirable characteristics, the fluorocarbon families of refrigerant were developed. From carbon tetrachloride R 12 was developed. Carbon "tet" as it is known in the industry is used as a

cleaning solvent. Because it contains no moisture it is ideal for cleaning compressor parts and in flushing a system.

Refrigerants used in mechanical refrigeration systems absorb heat from the area to be cooled and releases this heat outside of the cooled or refrigerated area. Because of the low boiling point of the refrigerant, it will vaporize in the system. In changing from a liquid state to a vapor it absorbs heat. The saturated vapor, so called because it is saturated with heat, will then flow into the compressor through the suction line. The compressor will compress the saturated vapor. It is then forced into the condenser. Here the heat is removed and the refrigerant returns to its liquid state as a high pressure liquid to repeat the cycle. Repeat cycles will not change the chemical characteristics nor decompose the refrigerant which is an important requirement of a refrigerant. The fluorocarbon refrigerants meet this requirement.

The most prominent refrigerants in use today are part of the fluorocarbon family of refrigerants. They are R 12 and R 22. R 500 an azeotropic refrigerant, so called because it is a combination of two refrigerants. R 500 was developed as a substitute for R 12, and has a greater refrigeration capacity then R 12 for comparable temperature and displacement. The cooling capacity of a unit will increase approximately 15% when R 12 is replaced with R 500. Because R 12 has a low condensing pressure it is ideal for use in automotive air conditioning. Because R 12 will not burn or explode it was adapted by the automotive air conditioners. R 12 is as inflammable as the carbon tet it was developed from.

To develop R 12 from carbon tet, two of the chlorine atoms was replaced with two fluorine atoms. The resulting compound was dichlorodifluoromethane, better known as R 12 and bearing the chemical symbol of $CC1_2F_2$. Because of its high stability and low operating pressures and temperatures R 12 is an ideal refrigerant. Another refrigerant developed for use in low temperature applications and air conditioning is R 22.

In domestic and commercial refrigeration R 12 is the most suitable refrigerant to use. Because operating pressures as well as temperatures are low, systems using R 12 do not require massive or heavy tubing or compressors as do ammonia systems. Looking at the temperature -pressure scale (*Figure 145*) it can be noted that the pressures are almost equal to the temperatures from 20 to 70°F. Another reason for R 12 to be suited for automotive air conditioning.

VACUUM IN BOLD TYPE			
F.	C.	R-12	R-22
—35	—37.23	**8.3**	2.7
—30	—34.44	**5.5**	5.0
—25	—31.67	**2.3**	7.5
—20	—28.89	**0.6**	10.03
—15	—26.12	2.4	13.3
—10	—23.33	4.5	16.6
—5	—20.56	6.8	20.3
0	—17.78	9.2	24.1
5	—15.00	11.8	28.3
10	—12.22	14.7	33.0
15	— 9.44	17.7	37.7
20	— 6.67	21.1	43.3
25	— 3.89	24.6	49.0
30	— 1.11	28.5	55.2
32	0.00	30.1	57.8
35	1.67	32.6	62.0
40	4.44	37.0	69.0
45	7.22	41.7	77.0
50	10.00	46.7	84.7
55	12.78	52.0	93.2
60	15.56	57.7	102.5
65	18.33	63.7	112.00
70	21.11	70.1	122.5
75	23.89	76.9	133.8
80	26.67	84.1	145.0
85	29.44	91.7	158.0
90	32.22	99.6	170.1

Comparable Temp. and Pressure

Figure 145 — Temperature Pressure Chart

R 12 can be considered a multi purpose refrigerant because its applciation is far and wide and covers most of the commercial and domestic requirements. It is used mainly in reciprocating compressors and some rotary types. R 12 uses are many, from household refrigeration through industrial refrigeration. R 12 is known to be a very versatile refrigerant in its limitless applications. R 12 is sometimes used in centrifugal compressors as a secondary coolant in the low stage of a cascade system.

R 22 another versatile refrigerant operates at a higher system pressure, but in exchange has lower compressor displacement requirements. Its uses are in most packaged air conditioners like room and window air conditioners, and is popular in central cooling units and heat pumps. It is used in freezers both domestic and commercially. Because of its low temperature application R 22 is also selected for commercial and industrial use.

R 500 is an azeotrope and has a slightly higher vapor pressure then R 12. R 500 will provide a higher capacity from the same compressor displacement presently using R 12. R 502 is also an azeotrope and its application is toward low evaporator temperatures. R 503 is an azeotrope and is used in the low stage of cascade systems. Compressor capacity is increased and it is suitable for low temperature application. R 12/31 is an axeotrope made up of R 12 and R 31. It is similar to R 12 but has some operating advantages. Like R 500, R 12/31 can be substituded for R 12 in existing equipment. A few of the advantages found in R 12/31 are increased cooling capacity, less power per BTU, improved heat transfer in the evaporator and condenser and compressor discharge temperatures are similar to R 12.

Besides the many other demands required of a quality refrigerant it is important that the refrigerant has a well balanced latent heat of vaporazation by comparison to the net refrigerating effect. This is measured in BTU/lb. The wider the ratio, the more efficient the refrigerant. In *Figure 146.* is illustrated the comparative figures for some of the fluorocarbon refrigerants.

Chemical Formula	R 12 $CC1_2F_2$	R 22 $CHC1F_2$	R 500 $CC1_2F_2/CH_3CHF_2$	Ammonia MH_3
* Boiling C^O	—29.79	—40.75	—33.5	—33.5
* Point F^O	—21.62	—41.36	—28.3	—28.0
* Freezing C^O	—158	—160	—159	—78.0
Point F^O	—252	—256	—254	—107.9

* Temperature at 1 atm

Figure 146 — Boiling Point of Some Refrigerants

In comparing R 12 and R 500 it can be noted that although for R 12 the latent heat of vaporization is 68.2 and the net refrigerating effect is 50.0, measured in BTU/lb. a percentage ratio of 26.68% exists between the two figures. In comparing this with R 500 which has a latent heat of vaporization of 82.5 and a net refrigerating effect of 60.6, measured in BTU/lb. it has a very close comparative percentage rate of 26.5%.

Refrigerant used in mechanical refrigeration system today, must meet rigid safety standards. Besides having all of the qualifications to use as a refrigerant it also must be non poisonous, non explosive and noncorrosive. They must be inflammable and readily detectable to locate leaks. The natural gas used for cooking purpose has a chemical added to if for the purpose of detecting, through the sense of smell, the odor of leaking gas. Adding such a chemical to the fluorocarbons would not be practical, because the stability of the refrigerant would be altered, and a leak within a refrigerator would contaminate the food. Fluorocarbon refrigerants will not cause food spoilage in itself. If a refrigerant leak occurs and the food is not refrigerated for a period of time the food will spoil. Leaks in a refrigeration system can be detected with the use of a halide torch or one of the many halogen leak detectors available to the trade.

The fluorocarbon family of refrigerants consists of two groups, those of the methane group such as R 11, R 12, R 13, R 14, R 21 and R 22. The other group known as the ethane members of the family are R 113 and R 114. A third group are called azeotropic refrigerants. Any refrigerant that is composed of two or more refrigerants is an azeotrope. By combining two or more mixtures of liquids a single compound is created known as an azeotrope. An important characteristic of such a compound is its stability. The boiling point remains constant during evaporation of the liquid, a most essential characteristic demanded of a refrigerant. In mixing liquids the thermal values take a radical change. It would be possible by error to make your own azeotropic refrigerant just by charging a system that may be low in refrigerant with the wrong refrigerant. The system may be improved, on the other hand, and most important of all, how would it effect the efficiency of the system.

R 500 is one of the most popular azeotropic refrigerants. It is composed of 73.8 percent of R 12 and 26.2 percent of R 153a. Pecentage by weight. It has a low boiling point of -28.3°F (-33.5°C.) R 500 was developed for use in small or medium-sized refrigeration applications such as medical therapy units. It can replace R 12 in existing equipment. By comparison to R 500, another of the azeotropes, R 502 has a lower boiling point of -49.*°F. (-45.4C.). Its refrigeration capabilities are greater than R 22. Its discharge temperature, which is the temperature of the refrigerant as it is being discharged through the high sde of the compressor to the condensor is equal to R 12. Its application in heat pumps is rapidly developing.

R 503, another azeotrope refrigerant, was developed for use in ultra low temperature applications such as cascade systems. It has an extremely low boiling point of -127.6°F. (-88.7°C.), compared to R 500 and R 502 In *Figure 146* illustrates the boiling points of some refrigerants.

To qualify as a desirable refrigerant for use in present day equipment, the refrigerant must have certain

qualities and properties inherent in the compound. Beside the necessity of a low boiling point, the refrigerants freezing point must be considerably below the temperature demands of the application.

Water could be a refrigerant, but it has a high boiling point at 212°F. Also at 32°F (0°C) water will freeze. It would take many more British Thermal Units (BTU) to evaporate water, and it could presumably freeze up in the system when the temperature of 32°F is attained. Desirable refrigerants have a low boiling point which enable the liquid to evaporate faster, and has a low freezing point, far below the temperature requirements of the equipment. *Figure 146* is a chart indicating the freezing and boiling points of the more popular refrigerants.

Figure 146 illustrates all fluorocarbon refrigerants have one thing in common, a low boiling point. The lower the boiling point, the faster the vaporization. The refrigeration cycle is based upon the ability of a refrigerant to evaporate and be reclaimed using the compressor and the condenser. The boiling point of a refrigerant generally indicates its best use. The temperature demand in relation to the boiling point helps to determine the right refrigerant for a particular use. Compounds substituting fluorine for hydrogen have higher molecular weights and quite often lower or extremely low boiling point. The fluorocarbon refrigerants, R 12, R 22 and R 500, have a rather high density but a low boiling point. They also have a lower viscosity and surface tension. The higher molecular weight is a contributing factor towards lower vapor specific heat values and lower latent heat of vaporization.

Pressure has a direct affect on the boiling points of refrigerants. When confined in a drum the liquid boils until it reaches a relative pressure. Releasing this pressure the refrigerant will evaporate. As evaporization takes place the tank will become cold. *Figure 147*, illustrates this process.

CRYOGENIC REFRIGERANTS

All refrigerants are not reclaimed and recycled however. Cryogenic refrigerants are a refrigerant type that is expandable. As illustrated in **Figure 147** the refrigerant is allowed to evaporate, and heat will be absorbed. The refrigerant vapors will become saturated. This will result in a lowered temperature of the refrigerant container and the immediate surrounding area. Because of their extremely low boiling point and high pressures it is impractible to condense

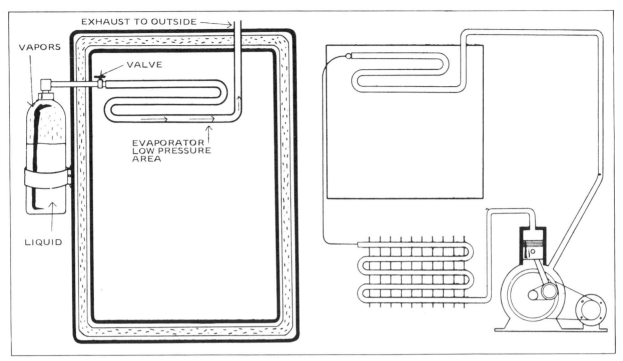

Figure 147 — Two Methods of Lowering Pressure

a cryogenic refrigerant. Used as cryogenic refrigerants are helium, hydrogen, nitrogen, oxygen and argon. These are all inert gases, helium having the lowest boiling point of all. Cyrogenic containers, equipped with pressure relief valves are constructed of special heavy duty materials to withstand the enormous pressures. These refrigerants must be stored in a cool place. Hydrogen especially is hgihly inflamable and extreme care should be taken against flames or sparks when in a hydrogen atmosphere.

If the liquid refrigerant is boiling at a higher temperature the pressure of the heat saturated vapor released from the refrigerant will also be higher. If the tempeature of the saturated vapor is higher, the condenser pressure will be higher as the heat is being removed. This holds true to a point. As the temperature of the saturated vapor rises, a point is reached where the vapors will no longer condense regardless of the increase of pressure. This temperature is called the critical temperature and its counterpart is called the critical pressure. *Figures 148* and *149* show the critical pressure and temperatures of the most common fluorocarbon refrigerants.

The fluorocarbon refrigerants do not burn, nor will they explode. They are non toxic, but under certain conditions can be hazardous. Fluorocarbon refrigerants released in the air, should be vented in

CRITICAL TEMPERATURES OF SOME FLUOROCARBONS	
R 12	233.6°F.
R 22	204.8°F.
R 502	179.9°F.
R 500	221.9°F.

Figure 148

CRITICAL PRESSURES OF SOME FLUOROCARBONS	
	(psia)
R 12	582.0
R 22	716.0
R 502	576
R 500	627.0

Figure 149.

closed unventilated areas, and must never be allowed to mix with fumes of other gases. The vapors of fluorcarbon refrigerant when mixed with fumes of an acetylene or butane torch produces a toxic gas. Never work with fluorocarbon refrigerants in an area where propane or butane flames are burning.

Allowing a drum or can containing refrigerant to be the direct rays of the sun may cause the container to explode. Always turn off all open flames such as range burners, water heaters, and gas furnaces including the pilot lights when purging refrigerants. Fluorocarbon refrigerants are not known to have an accumulative affect on the human body. However, breathing the fumes for a prolonged period of time can cause light headiness. Remaining in the environment of fluorocarbon fumes for too long a period may cause intoxication, and unconsciousness which can prove fatal. For this reason the work area must always be well ventilated when handling refrigerants.

A refrigerant must have the ability to withstand various temperature and pressure changes. Changing from vapor to liquid and from liquid to gas and retain its original characteristics is known as stability. The refrigerants capability to readily mix with oil is called miscibility. When a refrigerant is miscible with oil it should not affect the chemical stability of the refrigera nor should the refrigerant affect the lubricating qualities of the oil. This describes the characteristics of the fluorocarbon refrigerants. The fluorocarbon refirgerants meet the demands of the industry. The refrigerant will not decompose unless the refrigerant is exposed to extreme conditions. Moisture introduced into a system, either by chance or accident will trigger off a chain of events that may cause decomposure of the refrigerant, and corrosion of the system.

The hydrolisis rate (a chemical reaction in which the compound will react to ions of water to produce a weak solution of acid) for the fluorocarbon refrigerants is relatively lower than halogenated compounds. There are some differences within the group. Besides the temperatures and pressures the hydrolisis rate is affected by other materials in use. Moisture present in a system charged with a fluorocarbon refrigerant will, under certain conditions, form hydrochloride acid which is corrosive to metals. Moisture in an ammonia system will form ammonia hydroxide, which is also corrosive to metals. The condition can be further compounded by extreme heat and pressures. Carelessness in the handling of refrigerants is perhaps one of the major contributing factors in the decomposure of refrigerants. Mishandling tools and charging equipment add to this responsibility. Manifold gauges should be sealed with at least five pounds of pressure when not in use. (See page *125*, Strainers and Dryers).

Equal in importance of the refrigerants stability, is its miscibility with oil. Because it is necessary to lubricate parts of the compressor, such as bearings, valves, pistons cyclinder walls and related parts, the refrigerant is exposed to the lubricant. The oil must readily mix with the refrigerant and yet not disturb the lubricating qualities of the oil. Many of the fluoro-

carbon refrigerants are completely miscible with oil, while others are miscible to a degree. Under normal condenser pressures R 22 is miscible with oil. When oil enters the temperatures of the cold evaporator charged with R 22, the oil will tend to separate. Mixing with refrigerant will thin out the oil. Therefore a heavier viscosity of oil is used when the refrigerant is miscible with oil. Hermetic systems have a shaft seal which prevents large amounts of oil from entering the system, however there will be enough oil circulating in the system to lubricate the working parts of the compressor.

OIL AND LUBRICATION

A belt driven or direct couple compressor usually is lubricated by the "Splash system" and a good portion of the oil will escape to the system. Large tonnage units are equipped with oil traps *Figure 150* with a return line to the compressor. When a miscible refrigerant is employed lubrication is accomplished by the mixing of refrigerant and oil. In a sealed unit the oil is continually washing over the electrical windings of the motor. The physical characterstics of the oil should be such that the oil or motor are not affected.

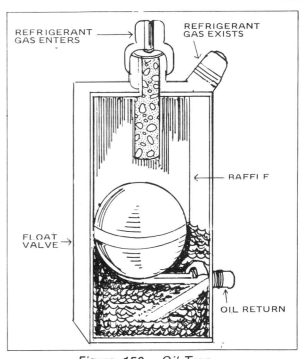

Figure 150 — Oil Trap

Fluidity of oil or its ability to flow fast or slow is expressed in viscocity. The viscocity of the oil must be compatible with the refrigerant, *Figure 151*. Some of the oil will circulate with the refrigerant, it is returned to the compressor in much the same way. When purging the refrigerant from a system in need of

REFRIGERANT	CHEM. SYMBOL	TRADE OR BRAND NAME	BOILING POINT F° AT (ATM P)	VISCOSITY RECOMMENDED
R 717	NH_3	Ammonia	−28.0	150-300
R 11 Trichloro- monofluoro- methane	CCl_3F	Carrene # 2 Freon-11 Frigen-11 Genetron-11 Isotron-11 Ucon-11	74.87	300
R 22 Monochloro- difluoro- methane	$CHClF_2$	Freon-22 Genetron-22 Isotron-22 Ucon-22	−41.36	150-300
R 502	$CHClF_2$ & C_2ClF_5	Freon-502	−49.8	150-300

Figure 151 — Recommended viscosity for Popular Refrigerants

repairs always purge into a glass container and secure both the container and hose to prevent tipping or breaking of the glass container. The hose should be secured inside the container to prevent the hose from blowing out. In this way any oil that is lost in the purging can be replaced in a like amount before recharging the system.

SPECIAL PURPOSE REFRIGERANTS

Ammonia, one of the oldest refrigerants known, also qualifies as an excellent refrigerant because of its low boiling point and the added advantage of low evaporation and condensing pressures. This added bonus makes in unnecessary for the equipment to be massive and heavy. Ammonia, given the assigned number of R 717, is one of the special purpose refrigerants. Its use in low temperatures at relative low condenser pressure makes ammonia a desirable refrigerant. Ammonia is used in ice cream plants, and for cold storage. Many ice houses use ammonia equipment. The boiling point of ammonia is -28°F, and evaporation will occur at low pressure. Ammonia systems are water cooled because of extreme high discharge temperatures. Ammonia is lighter then air and colorless. The refrigerant is chemically staple. Ammonia will not decompose nor affect the lubricating qualities of oil. Introduction of mositure into an ammonia system will cause the oil to immulsify.

Ammonia when mixed with water will form ammonia hydroxide, a very strong alkali solution that will attack nonferrous metals. Copper and brass are never used in ammonia systems. Ammonia will burn when exposed to an open flame. Compressed with air in the right proportions can cause ammonia to explode. Ammonia fumes are irritating to the mucous membrane and the eyes.

Leaks in an ammonia system can be detected with a burning sulfer candle. A dense white smoke will form when the fumes are mixed. In this manner a leak can be pinpointed.

R 502, one of the azeotropic refrigerants, was developed as a special purpose refrigerant. Composed of 48.8 percent of R 22 and 51.2 percent of R 115, it has a low boiling point of -49.76°F. This low boiling point makes it applicable to low temperature refrigeration. Its refrigeration capacity is greater then R 22, and discharge temperatures of R 502 are comparable to R 12. It is primarily used in low and medium refrigeration applications, such as commercial freezers and display cases. The low boiling point of R 502 also make it an excellent refrigerant for use in air conditioning and heat pumps.

Another special purpose refrigerant is R 114. R 114 has a relatively high boiling point at 38.8°F. R 114 is generally used in small refrigeration equipped with rotary type compressors. It is also used in industrial process cooling and air conditioning systems equipped with multistage centrifugal compressors.

Carbon dioxide one of the few compounds that will change from a solid state to a vapor without first turning to a liquid, is classified as a special purpose refrigerant. Dry ice a substance most people are familiar with is frozen carbon dioxide, the chemical symbol is CO_2. Carbon dioxide boils at -109°F. Because evaporator pressures will exceed 300 psi and condenser pressures are above 1000 psi it is necessary to use heavy and massive refrigeration equipment with corrospondingly heavy piping. It is non toxic and is used as an additive in soda pop. It is inflammable and used in certain type of fire extinguishers.

NOTES

GEM PRODUCTS, INC. MASTER PUBLICATIONS | APPLIANCE REPAIR MANUALS

REPAIR-MASTER® For Automatic Washers, Dryers & Dishwashers

These Repair-Masters® offer a quick and handy reference for the diagnosis and correction of service problems encountered on home appliances. They contain many representative illustrations, diagrams and photographs to clearly show the various components and their service procedure.

Diagnosis and repair charts provide step-by-step detailed procedures and instruction to solve the most intricate problems encountered in the repair of washers, dryers and dishwashers.

These problems range from timer calibrations to complete transmission repair, all of which are defined and explained in easy to read terms.

The Repair-Masters® are continually updated to note the latest changes or modifications in design or original parts. These changes and modifications are explained in their service context to keep the serviceman abreast of the latest developments in the industry.

AUTOMATIC WASHERS
9001 Whirlpool
9003 General Electric
9005 Westinghouse Front Loading
9009 Frigidaire
9010 Maytag
9011 Philco-Bendix Front Loading
9012 Speed Queen
9015 Norge – Plus Capacity
9016 Frigidaire Roller-Matic
9017 Westinghouse Top Loading

DISHWASHERS
5552 General Electric
5553 Kitchenaid
5554 Westinghouse
5555 D & M
5556 Whirlpool
5557 Frigidare

CLOTHES DRYERS
8051 Whirlpool-Kenmore
8052 General Electric
8053 Hamilton
8055 Maytag
8056 Westinghouse
8057 Speed Queen
8058 Franklin
8059 Frigidaire

*The D & M Dishwasher Repair-Masters covers the following brands: Admiral — Caloric — Chambers — Frigidaire (Portable) — Gaffers and Sattler — Gibson — Kelvinator — Kenmore — Magic Chef — Magic Maid — Norge — Philco — Pioneer — Preway — Roper — Wedgewood — Westinghouse (Portable)

The Repair-Masters® series will take the guesswork out of service repairs.